Signals and Communication Technology

This series is devoted to fundamentals and applications of modern methods of signal processing and cutting-edge communication technologies. The main topics are information and signal theory, acoustical signal processing, image processing and multimedia systems, mobile and wireless communications, and computer and communication networks. Volumes in the series address researchers in academia and industrial R&D departments. The series is application-oriented. The level of presentation of each individual volume, however, depends on the subject and can range from practical to scientific.

Indexing: All books in "Signals and Communication Technology" are indexed by Scopus and zbMATH

For general information about this book series, comments or suggestions, please contact Mary James at mary.james@springer.com or Ramesh Nath Premnath at ramesh.premnath@springer.com.

More information about this series at http://www.springer.com/series/4748

Cristian Monea · Nicu Bizon

Signal Processing and Analysis Techniques for Nuclear Quadrupole Resonance Spectroscopy

Springer

Cristian Monea
Research and Development Department
Mira Technologies Group
Voluntari, Romania

Doctoral School of Electronics,
Telecommunications and Information
Technology
Politehnica University of Bucharest
Bucharest, Romania

Nicu Bizon
Faculty of Electronics, Communications
and Computer Science
University of Pitesti
Pitesti, Arges, Romania

Doctoral School of Electronics,
Telecommunications and Information
Technology
Politehnica University of Bucharest
Bucharest, Romania

ISSN 1860-4862 ISSN 1860-4870 (electronic)
Signals and Communication Technology
ISBN 978-3-030-87860-3 ISBN 978-3-030-87861-0 (eBook)
https://doi.org/10.1007/978-3-030-87861-0

This Springer imprint is published by the registered company Springer Nature Switzerland AG
The registered company address is: Gewerbestrasse 11, 6330 Cham, Switzerland

Foreword

This book brings new literature reviews on nuclear quadrupole resonance (NQR) techniques, appropriate signal processing techniques proposed for NQR spectrometers and implemented NQR spectrometers developed to date.

The signals obtained from NQR spectroscopy are simulated to propose an NQR signal model to be used on the comparative analysis (based on a new set of performance indicators proposed here) of the main signal processing methods that have been developed in the literature.

So, this book proposes a new approach to detecting the NQR signal by using artificial intelligence methods (such as deep learning techniques and other proposed advanced techniques). As a result of this study, the best performing solutions are selected, adapted for NQR detection and implemented in a customized NQR spectrometer and, finally, the NQR-based equipment was evaluated in real conditions.

The data set was acquired under different operating conditions and was analysed to evaluate the performance of the custom-designed NQR spectrometer.

It is worth mentioning that the implementation of the detection system is detailed in this book from both a hardware and a software point of view, and the initial deep learning technique proposed is further improved in the penultimate chapter by using transfer learning and autoencoders to allow its use in other spectrometers and at different noise levels.

Thus, the content of this book is interdisciplinary, involving knowledge of physics, electrical and electronics engineering, signal processing, signal analysis and artificial intelligence (machine learning, deep learning, etc.) and optimization.

Therefore, the content of this book is addressed to researchers studying the resonance of nuclear quadrupoles and its applications, and NQR equipment manufacturers and readers in the multidisciplinary fields mentioned above, who want to understand the NQR spectrometer and the techniques used in its design.

So, both researchers (physicists, engineers and computer scientists) and engineers working for manufacturers of security or laboratory equipment based on the NQR technique, as well as master's and doctoral students studying and researching NQR detection techniques, will all be interested to read this book because it proposes (i) a new and performant NQR signal detection techniques based on improved deep learning methods and (ii) a detailed description regarding the hardware and software implementation of an NQR detection equipment that implements the above-mentioned solutions.

In conclusion, it is noteworthy that the content of the chapters is presented gradually and theoretically in detail as necessary to understand the problems and effective techniques for implementing the NQR, being highly recommended for study in education and research.

Baku, Azerbaijan
April 2021

Arif M. Hashimov

Preface

In recent years, global terrorism and organized crime, especially in the field of drug trafficking, have increased enormously, with more than 10,000 incidents per year reported in the global terrorism database since 2017. For example, in 2017, over 140 tons of cocaine and 5 tons of heroin were confiscated in the European Union, and the number of confiscations has increased every year. In this context, a number of prohibited substance control systems have been developed and installed worldwide (at airports, customs, post offices and border checkpoints) to combat illegal activity and ensure the safety of citizens.

Therefore, this book addresses the issue of improving the detection of prohibited substances using the nuclear quadrupole resonance (NQR) technique and takes into account the advanced signal processing techniques recently proposed in the literature, including by the authors of this book.

NQR is a confirmed analysis technique and recognized as a solution that meets the performance standards imposed for security checkpoints. In short, the NQR technique is based on scanning the object (package and luggage) which may contain a substance of interest. An electromagnetic pulse train (called the excitation sequence) is used for scanning, and the response is received at the end of the excitation sequence. The NQR technique has the advantage of a high specificity relative to a substance of interest, but it is also worth mentioning that it is very difficult to detect the response (the weak signal emitted by the substance).

This book analyses and proposes new signal processing and analysis methods for the detection of dangerous or contraband substances (such as explosives, narcotics or toxic substances). Also, the hardware solutions implemented in a custom NQR spectrometer (detector) are described, and the performances obtained are compared with similar ones reported in the literature.

The identification of prohibited substances by nuclear quadrupole resonance depends on the physical principle used and deployment location of the NQR equipment, because the received signal has a low signal-to-noise ratio and is susceptible to radio interference in the same frequency band. In addition, the spectral lines are dependent on the substance's temperature.

Because the detection time is critical and the number of false-positive and false-negative alarms must be minimized, the main signal processing challenges are related to the ability to detect the response signal using a limited number of acquisitions, to perform real-time analysis of the response signal's characteristics in order to cancel radio frequency noise and compensate the temperature dependence of spectral lines, with the ultimate goal of maximizing the detection accuracy.

In the last decade, numerous laboratory NQR detection systems have been proposed and implemented for identifying prohibited substances, with detection probabilities of up to 90%. However, the detection performance decreases greatly in the operational environment due to the causes mentioned above. So, new hardware and software solutions are needed to increase the performance of NQR detection. Therefore, the purpose of this book is to present some hardware and software solutions to improve the performance of the NQR technique for detecting prohibited substances.

This book is structured in ten chapters, and Chap. 1 introduces the topic of nuclear quadrupole resonance detection of prohibited substances.

Chapter 2 briefly presents the methods and equipment for the acquisition and analysis of signals for detection of prohibited substances.

Chapter 3 continues the analysis started in Chap. 2, focusing on nuclear quadrupole resonance spectroscopy and presenting its physical principle, specific signal acquisition methods, and laboratory and commercial NQR equipment.

Chapter 4 presents in detail the specific pre-processing, post-processing and signal analysis techniques proposed for nuclear quadrupole resonance detection.

Chapter 5 presents the test signals that will be used in the following chapter to study the performance of the algorithms proposed in the literature for NQR signal processing and analysis.

The study of the performance of the NQR algorithms is presented in Chap. 6 in order to define a reference regarding the obtained performance. It considers specific performance indicators and a multi-criteria classification algorithm based on an objective function to improve the probability of detection.

The experimental results of the acquisition of nuclear quadrupole resonance signals are presented and analysed in Chap. 7 in order to identify new and improved NQR detection techniques.

Thus, Chap. 8 presents new signal analysis algorithms based on machine learning and demonstrates their improved performance in detecting substances using the NQR technique compared to those presented in the previous chapter.

Chapter 9 describes techniques for enhancing the performance of the machine learning algorithm proposed in the previous chapter that are based on transfer learning and noise reduction using autoencoders.

The last chapter presents the implementation of an NQR equipment used for signal acquisition, processing and analysis. The hardware and software implementations are detailed, and the system's performance in different operating conditions is evaluated.

Summarizing, the salient features of the book would be the following:

- Presenting a detailed literature review of the methods and equipment for signal acquisition and analysis for the detection of prohibited substances,
- Presenting the principle of nuclear quadrupole resonance spectroscopy,

- Providing a detailed literature review of the signal processing and analysis techniques applied in nuclear quadrupole resonance detection, with a new classification,
- Performing a modeling of the signals used in nuclear quadrupole resonance spectroscopy, with the proposal of a new signal model,
- Performing a comparative study of the signal processing and analysis algorithms, in order to verify and create its own reference for the performance indicators, that will be of help for comparing the newly developed solutions,
- Performing an analysis of the signals acquired from nuclear quadrupole resonance experiments with the aim of aiding the development of new detection solutions,
- Proposing new signal analysis algorithms based on artificial intelligence/deep learning in order to improve the detection of substances by NQR technique,
- Proposing new features for the classification of NQR signals using artificial intelligence/deep learning algorithms,
- Proposing new artificial intelligence/deep learning techniques for enhancing the performance of signal analysis algorithms,
- Detailing the hardware and software implementations of an NQR signal processing and analysis system,
- Helpful for researchers and practitioners in the areas of electrical engineering, signal processing and analysis, applied spectroscopy, as well as for security or laboratory equipment manufacturers.

In conclusion, electrical engineering students and specialists are eligible and guided to use the book effectively in their current activities.

Pitesti, Romania
April 2021

Cristian Monea
Nicu Bizon

Contents

Abbreviations

ACC	Accuracy
AE	Autoencoder
AI	Artificial Intelligence
ALE	Adaptive Line Enhancement/Adaptive Line Enhancer
ALS	Alternating Least Squares
AML	Approximate Maximum Likelihood
ANC	Adaptive Noise Cancellation/Adaptive Noise Canceller
APES	Amplitude and Phase Estimator
API	Application Program Interface
AR	Autoregressive
ARMA	Autoregressive Moving Average
AU	Arbitrary Unit
AUC	Area Under Curve
BICETAML	Beamforming-based Interference Cancellation Echo Train Approximate Maximum Likelihood
CAPA	Capon Amplitude and Phase Estimator Average
CAPES	Capon Amplitude and Phase Estimator
CMA-ES	Covariance Matrix Adaptation Evolution Strategy
CNN	Convolutional Neural Network
CNN-LSTM	Convolutional Neural Network Long Short-Term Memory
cNQR	Classical NQR
ConvAE	Convolutional Autoencoder
ConvLSTM	Convolutional Long Short-Term Memory
CPMG	Carr–Purcell–Meiboom–Gill
CS	Clonal Selection
c-SEAQUER	Conventional-Subspace-based EvaluAtion of Quadrupole resonance signals Exploiting Robust methods
CTLMS	Continuous Time Least Mean Squares
CUDA	Compute Unified Device Architecture
CW	Continuous Wave
DB	Database
DBN	Deep Belief Network

DC	Direct Current
DCRCB-GLRT	Doubly Constrained Robust Capon Beamformer-Generalized Likelihood Ratio Test
DEPIC	Dual-channel Estimation of Phase and amplitude for Interference Cancellation
DFT	Discrete Fourier Transform
DL	Deep Learning
DLL	Dynamic-Link Library
DMA	Demodulated Approach
DMNB	2,3-Dimethyl-2,3-dinitrobutane
DNN	Deep Neural Network
DNN-AE	Deep Neural Network Autoencoder
EC	Ensemble of Classifiers
EFG	Electric Field Gradient
EGDN	Ethylene Glycol Dinitrate
EI	Expected Improvement
EPIC	Estimation of Phase and amplitude for Interference Cancellation
ESPIRE	Exploiting Spatial diversity and Polymorphism In Robust Estimation (of NQR signals)
ESPRIT	Estimation of Signal Parameters via Rotational Invariance Technique
ET	Echo Train/Extra Trees
ETAML	Echo Train Approximate Maximum Likelihood
ETCAPA	Echo Train Capon APES Average
ET-ESPRIT	Echo Train-Estimation of Signal Parameters via Rotational Invariance Technique
FDR	False Discovery Rate
FETAML	Frequency-selective Echo Train Approximate Maximum Likelihood
FFT	Fast Fourier Transform
FHETAML	Frequency-selective Robust Echo Train Approximate Maximum Likelihood
FICAML	Fourier analysis-based Interference Cancellation Approximate Maximum Likelihood
FID	Free Induction Decay
FLSETAML	Frequency-selective Least Squares Echo Train Approximate Maximum Likelihood
FN	False Negative
FNR	False-Negative Rate
FODETAML	Frequency-selective Offset-Dependent Echo Train Approximate Maximum Likelihood
FOR	False Omission Rate
FP	False Positive
FPGA	Field-Programmable Gate Array

FPR	False-Positive Rate
FRETAML	Frequency-selective Robust Echo Train Approximate Maximum Likelihood
FSAML	Frequency-Selective Approximate Maximum Likelihood
FSMC	Frequency-Selective Multichannel detector
FT-NQR	Fourier Transform NQR
GA	Genetic Algorithm
GBT	Gradient Boosted Trees
GDDR	Graphics Double Data Rate
GP	Gaussian Process
GPIO	General-Purpose Input/Output
HETAML	Hybrid Robust Echo Train Approximate Maximum Likelihood
HMTD	Hexamethylene Triperoxide Diamine
HMX	High Melting Explosive/Her Majesty's Explosive/High-velocity Military Explosive/High-Molecular-weight RDX
I2C	Inter-Integrated Circuit
ICAML	Interference Cancellation Approximate Maximum Likelihood
ICETAML	Interference Cancellation Echo Train Approximate Maximum Likelihood
ICFETAML	Interference Cancellation Frequency-selective Echo Train Approximate Maximum Likelihood
IDE	Integrated Development Environment
ILSVRC	ImageNet Large-Scale Visual Recognition Challenge
IMS	Ion Mobility Spectrometry
ISR	Interference-to-Signal Ratio
JSON	JavaScript Object Notation
k-NN	k-Nearest Neighbour
LMS	Least Mean Squares
LNA	Low-Noise Amplifier
LS	Least Squares
LSETAML	Least Squares Echo Train Approximate Maximum Likelihood
LS-FRETAML	Least Squares-Frequency-selective Robust Echo Train Approximate Maximum Likelihood
LSTM	Long Short-Term Memory
M3L	Model-Mismatched Maximum Likelihood
MDMA	3,4-Methylenedioxymethamphetamine (Ecstasy)
MFD	Multi-criteria Fourier-based Decision algorithm
ML	Machine Learning
MLH	Maximum Likelihood
MSWF	Multi-Stage Wiener Filter
MUSIC	Multiple Signal Classification Spectral Estimation
NG	Nitroglycerin
NLS	Nonlinear Least Squares
NMOS	Negative-channel MOS
NMR	Nuclear Magnetic Resonance

NN	Neural Network
NORRDIQ	Noise and RFI Removal for Detection In QR applications
NPAPS	Non-Phase-Alternated Pulse Sequence
NPV	Negative Predictive Value
NQR	Nuclear Quadrupole Resonance
NS	Negative Scan
ODETAML	Offset-Dependent Echo Train Approximate Maximum Likelihood
PAPS	Phase-Alternated Pulse Sequence
PASLSE	Phase-Alternated Spin-Locked Spin Echo
PCC	Pearson Correlation Coefficient
PETN	Pentaerythritol Tetranitrate
PIN	PIN diode
PNP	PNP bipolar transistor
PostPA	Post-Processing and Analysis
PPV	Positive Predictive Value
PS	Positive Scan
PSL	Pulsed Spin Locking
QCPMG	Quadrupolar Carr–Purcell–Meiboom–Gill
RAM	Random Access Memory
RCB	Robust Capon Beamformer
RCDAML	Robust Correlation Domain Approximate Maximum Likelihood
RDX	Hexogen
ReLU	Rectified Linear Unit
REMIQS	Robust Estimation of MultIple polymorph QR Signals
ResNet-34	Residual neural Network-34
RESPEQ	Robust Evaluation using Subspace-based methods of Polymorphic nuclEar Quadrupole signals
RETAML	Robust Echo Train Approximate Maximum Likelihood
REWEAL	Robust Evaluation With conic subspace constraints of Estimation of Approximate maximum Likelihood
RF	Radiofrequency/Random Forest
RFI	Radio Frequency Interference
RLS	Recursive Least Squares
RMS	Root Mean Square
RNN	Recurrent Neural Network
ROC	Receiver Operating Characteristic
RTDAML	Robust Time Domain Approximate Maximum Likelihood
RX	Reception
SA	Simulated Annealing
SAMP	Spectral Average to Maximum Peak
SCB	Standard Capon Beamformer
SE	Spin Echo

SEAQUER	Subspace-based EvaluAtion of Quadrupole resonance signals Exploiting Robust methods
SF	Spectrometer Frequency
SLMP	Spin-Locked Multi-Pulse
SLSE	Spin-Locked Spin Echo
SMBO	Sequential Model-Based Optimization
SMF	Spectral Maximum Frequency
sNQR	Stochastic NQR
SNR	Signal-to-Noise Ratio
SOI	Signal Of Interest
SORC	Strong Off-Resonance Comb
SPA	Signal Processing and Analysis
SPI	Serial Peripheral Interface
SQUID	Superconducting Quantum Interference Device
SRAR	Spectral Regions Average Ratio
SRO	Super-Regenerative Oscillator
SSD	Solid-State Drive
SSFP	Steady-State Free Precession
SVM	Support Vector Machine
SW	Spectral Width
TATP	Triacetone Triperoxide
THC	Tetrahydrocannabinol
TN	True Negative
TNR	True-Negative Rate
TNT	Trinitrotoluene
TP	True Positive
TPE	Tree Parzen Estimator
TPR	True-Positive Rate
TX	Transmission
UART	Universal Asynchronous Receiver/Transmitter
USB	Universal Serial Bus
VGG-16	Visual Geometry Group-16
WICAML	Wavelets-based Interference Cancellation Approximate Maximum Likelihood
WPF	Windows Presentation Foundation
WURST-QCPMG	Wideband Uniform Rate Smooth Truncation-Quadrupolar Carr–Purcell–Meiboom–Gill
XAML	Extensible Application Markup Language

Chapter 1
Introduction

The last decades have seen an unprecedented increase in global terrorism and organized crime, especially in the area of drug trafficking. The global terrorism database indicates over 10,000 incidents in 2017 alone [1], and recent events include the bombings from Sri Lanka, 21 April 2019, Kabul, 19 August 2019 or Beirut, August 4, 2020. In 2017, more than 140 tons of cocaine and 5 tons of heroin were reported to be seized in the European Union [2]. Figures 1.1 and 1.2 illustrate the evolution of heroin trafficking and terrorism at European and global level. In this context, a number of prohibited substance control systems have been developed and installed in airports, customs, post offices and border checkpoints to combat illegal activity and increase the safety of citizens.

The detection techniques applied range from trace detection, based on ion mobility spectrometry, chemiluminescence, colorimetric tests or mass spectrometry, to the detection of masses of substances, based on imaging techniques (X-rays, millimeter waves, computed tomography or terahertz waves) and nuclear techniques (thermal/fast neutrons, nuclear magnetic resonance or quadrupolar nuclear resonance). Also, there are various methods of signal processing and analysis, depending on the detection technique applied. They include statistical algorithms for signal detection and estimation, spectral processing, adaptive filtering, dedicated image processing algorithms or advanced signal analysis and recognition techniques based on machine learning (ML).

Nuclear Quadrupole Resonance (NQR) has been acknowledged as a technique for scanning prohibited substances that can meet the high requirements of security checkpoints. It has also been applied in landmine detection and its use is justified by the fact that there are currently over 100 million active landmines worldwide, killing or injuring, on average, 5,000 people annually [4].

C. Monea and N. Bizon, *Signal Processing and Analysis Techniques for Nuclear Quadrupole Resonance Spectroscopy*, Signals and Communication Technology,
https://doi.org/10.1007/978-3-030-87861-0_1

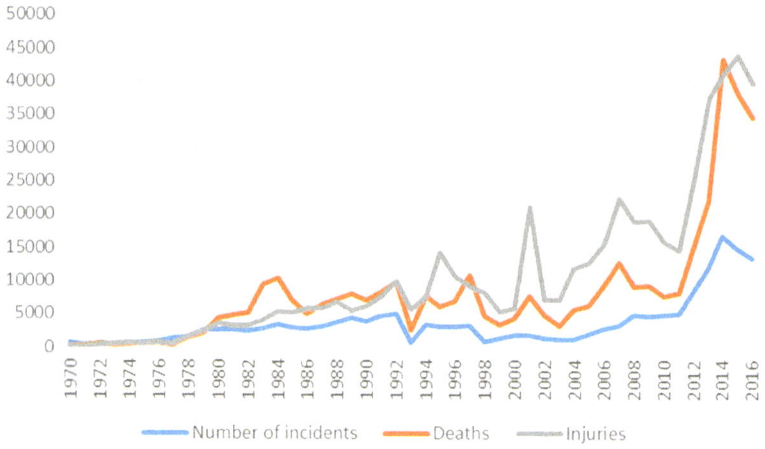

Fig. 1.1 The evolution of worldwide terrorist incidents during 1970–2016 (in total, 182,438 incidents) [3]

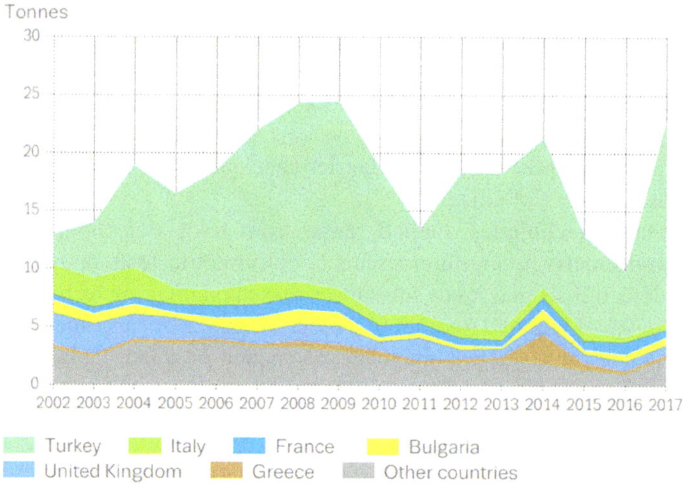

Fig. 1.2 The quantity of heroin captured in the European Union during 2002–2017 [2] (© European Monitoring Centre for Drugs and Drug Addiction, 2019)

1.1 Context of This Work

The detection of prohibited substances (explosives, drugs, toxic substances) occupies an important place in the field of security. Therefore, a wide range of techniques and equipment has been developed for the identification of landmines or narcotics and explosives hidden in luggage or clothing.

Generally, the techniques used for the detection of substances are classified according to the target sought, in techniques for bulk and trace detection. The first aims to detect a mass of substance which has certain properties that identify it as explosive, narcotic or toxic or, in other words, seek the detection of the substance itself. The techniques of the second category aim to identify the presence of an explosive compound based on the vapors emitted by it or the explosive particles stored on nearby surfaces. NQR is a technique for bulk detection that has several advantages over others.

NQR spectroscopy is a solid-state analysis technique that is based on the excitation of the scanned substance with a train of electromagnetic pulses and the reception of its response after the excitation ends. The frequency of the response signal is strongly influenced by the chemical structure of the substance, which gives the NQR technique a very high specificity, allowing the detection of a substance through its unique radio "fingerprint" [5]. The technique can be applied to substances that contain quadrupolar nuclei, such as the ^{14} N isotope found in most explosives, narcotics, drugs and toxic substances. Nuclear quadrupole resonance detection equipment offers several advantages over others, such as very high specificity, entire volume scanning using non-ionizing radiation, applicability to humans, while allowing non-invasive detection, and the fact that the detection accuracy does not depend on of the mass or form of the substance, instead it is determined by its chemical structure. The advantages of this technique have allowed its use in security applications for scanning luggage and people, but also the identification of counterfeit drugs.

From the point of view of signal processing and analysis applied in the field of detection of prohibited substances, there are numerous solutions that depend on the detection technique. One can mention the application of signal processing and estimation techniques for differentiating information-bearing signals from random ones and their quantitative evaluation, such as optimal filtering, unknown parameters estimation, power spectrum analysis or statistical algorithms for detection. Regarding nuclear quadrupole resonance, several techniques have been proposed to improve detection, based on statistical algorithms, spectral estimation techniques, noise and interference rejection solutions, adaptive filtering or classification. The first techniques aimed to perform detection by Fourier spectral analysis and evaluation of the peak signal response, but their low performance in low signal-to-noise ratio conditions has led to the research of more advanced techniques that can exploit signal characteristics such as the structure of the spectral lines or the response in the case of polymorphic substances and mixtures. Thus, statistical algorithms have been developed for filtering and canceling interference for both single-channel and multi-channel detection. A different approach has recently been proposed using artificial intelligence techniques for signal response analysis, such as neural networks, support vector machines and decision trees to increase the probability of detection, with the latter solution reaching a probability of 85.7%.

The challenges of nuclear quadrupole resonance detection of prohibited substances are related to the physical principle used, but also to the requirements imposed by the operational environment in which the equipment based on this technique is installed. NQR has the disadvantages that it has a low signal-to-noise ratio, is susceptible to the influence of radio frequency interference, and the spectral lines depend on the substance's temperature. Generally, it is desired to detect prohibited substances in the shortest possible time, with minimal risk of false alarms or incorrect identifications, with the highest possible specificity and sensitivity, while maintaining a low cost for the equipment. From the point of view of signal processing and analysis, the challenges of this technique are to detect the response signal under a low signal-to-noise ratio, using a limited number of acquisitions, exploiting the characteristics of the response signal, canceling interference and compensating for spectral lines' temperature dependence, with the ultimate goal of maximizing the probability of detection and minimizing the probability of false positive and false negative alarms.

1.2 Purpose of This Work

Numerous NQR detection systems for prohibited substances have been developed both at the laboratory level for landmine detection, counterfeit drug identification, luggage and person scanning, and at the commercial level for airports, post offices and security points, reaching detection probabilities of up to 90% [6]. However, existing systems have detection performance that is unacceptable in the operational environment. Specifically, unidentified landmines or undetected explosives in luggage or packages can lead to material damage and even human casualties. There is a clear need to increase the performance of NQR detection, both by improving the hardware of the equipment and by proposing better software solutions for processing and analyzing the response signal.

Therefore, the purpose of this work is to propose hardware and software solutions to improve the detection of prohibited substances by the nuclear quadrupole resonance technique.

From a hardware point of view, several solutions are being researched to reduce noise and compensate for the factors that influence NQR detection. The improvements brought by the proposed solutions are evaluated quantitatively and highlighted by comparison with other detection systems.

From a software point of view, the aim is to improve detection by applying pre-processing, processing and signal analysis techniques. Various pre-processing techniques are researched to increase the signal-to-noise ratio (SNR) as the response signal is acquired. Regarding signal processing and analysis, several deep learning (DL) algorithms are being studied which have provided very good results in classification applications. The algorithms are applied to NQR detection and are evaluated comparatively in order to select the most efficient one.

These solutions are implemented in a detection equipment that ensures the pre-processing, processing and analysis of the NQR response signal, intended for use in scanning points for illicit substances.

1.3 Structure of This Work

This paper is structured in eleven chapters. The **current chapter** has an introductory role and is intended to present the topic and structure of this work, whereas Chaps. 2–10 represent its main body. Each chapter contains a section of conclusions, and the lists of figures, tables and abbreviations, annexes and references are added to these.

Chapter 2 is a bibliographic study of the methods and equipment for the acquisition and analysis of signals for the detection of prohibited substances. The study starts from the general framework of signal detection, gradually focusing on the detection techniques of prohibited substances. The first section is an introduction to the theory of signal detection and estimation, as a general framework for the detection of prohibited substances. Then, the study focuses on commercial and laboratory techniques and equipment for the detection of prohibited substances, presenting its physical principles and conducting a comparative analysis of them in terms of advantages and disadvantages.

Chapter 3 continues the bibliographic study and focuses on nuclear quadrupole resonance spectroscopy. It presents the physical principle underlying NQR spectroscopy, the acquisition methods, and existing laboratory and commercial equipment.

Chapter 4 focuses on signal processing and analysis in nuclear quadrupole resonance detection, by carrying out a bibliographic study of the pre-processing, post-processing and signal analysis techniques. The first section presents the signal averaging technique and quadrature detection both mathematically and from an applicative point of view, highlighting the advantages and disadvantages of each. The second section presents the post-processing and analysis techniques used in NQR detection. For clarity, they are described in separate sub-sections as techniques for single-channel and multi-channel detection, respectively. Several categories are investigated, such as spectral estimation techniques, statistical solutions, techniques for radio frequency interference rejection, adaptive filtering or classification. The last section identifies the possibilities for further research in this field.

Chapter 5 models the signals used in nuclear quadrupole resonance spectroscopy. It presents the signal, noise and interference models used in NQR spectroscopy, and a new signal model is proposed that better approximates the actual shape of an echo. The signal modeling is carried out by simulations and the results are used in the following chapter for making a study of the signal processing and analysis algorithms applied in NQR detection.

Chapter 6 is a comparative study of the signal processing and analysis algorithms, in order to verify and create its own reference for the performance indicators. In the first section, the algorithms that will be studied are proposed and a brief presentation of them is made. In order to give a higher degree of novelty to the comparative analysis, a multi-criteria algorithm is proposed to improve the probability of detection. The last section defines the analysis procedure and the measured performance indicators, describes the implementation aspects of the simulation program, and finally presents and interprets the results obtained from the analysis of the investigated algorithms.

Chapter 7 presents an analysis of the signals acquired from nuclear quadrupole resonance experiments with the aim of aiding the development of new detection solutions. Thus, the first section presents the parameters and acquisition scenarios for obtaining the data set used to develop signal analysis algorithms. Then the signal characteristics that they will use are established and a descriptive statistical analysis of the data is carried out in order to obtain an overview and to observe the correlations between the different characteristics.

Chapter 8 proposes new signal analysis algorithms based on machine learning in order to improve the detection of substances by NQR technique. The first section proposes the algorithms that will be investigated and adapted to this type of detection. The second and third sections present implementation aspects of the algorithms and workflow and performs the training and comparative evaluation of algorithms. Finally, the algorithm identified as the most efficient is optimized, and the results obtained are compared with those in the previous chapter.

Chapter 9 proposes two solutions for improving detection that use the optimized algorithm: its generalization using transfer learning and noise reduction using autoencoder networks.

Chapter 10 presents the signal processing and analysis system developed for NQR detection. The first section deals with the implementation of the hardware detection system, presenting the general architecture of the spectrometer and the solutions to improve the detection. The second section describes the implementation of the system from a software point of view, starting from the general architecture of the signal control and processing software and describing its components. The last section deals with the evaluation of the detection system. In this sense, different characteristics of the its modules are evaluated such as the rejection factor of the radio frequency switch, the improvement brought by the ferrite core and the pre-processing techniques, but also the detection limits of the system.

References

1. Global Terrorism Database.: Incidents over time. Internet: www.start.umd.edu/gtd/search/Results.aspx?start_yearonly=1970&end_yearonly=2017. 10 Aug. 2020
2. European Monitoring Centre for Drugs and Drug Addiction.: European Drug Report: Trends and Developments (2019).
3. Wikipedia.: GTD. Internet: http://en.wikipedia.org/wiki/Global_Terrorism_Database. 10 Aug. 2020.
4. Minesweepers.: Facts about landmines. Internet: www.landminefree.org/2017/index.php/support/facts-about-landmines. 10 Aug. 2020.
5. Chihara, H. (ed).: Nuclear Quadrupole Resonance Spectroscopy Data. Landolt-Börnstein—Group III Condensed Matter (1997)
6. Hibbs, A.D.: Man portable mine detector using nuclear quadrupole resonance—first year progress and test results. In: Second International Conference on Detection of Abandoned Land Mines (1998)

Chapter 2
Methods and Equipment for Signal Acquisition and Analysis for the Detection of Prohibited Substances

2.1 Signal Detection and Estimation Theory

Signal detection theory, also called detection theory, is a field that aims to identify the occurrence of an event with a certain degree of certainty and extract information about it. The detection theory aims to differentiate information-carrying signals from random ones, called noise, caused by the environment or receiver. The signal estimation theory is complementary to the detection theory, and aims to quantitatively evaluate the event or information and measure the detection's degree of certainty.

The detection theory initially emerged in the military, during World War II, in response to the need to defend valuable resources by detecting and destroying enemy aircraft and missiles in all types of environmental conditions [1]. This field has advanced since the development of radar, sonar, digital communications and digital signal processing.

All signals are affected by some form of noise or uncertainty, which complicates the task of the receiver to extract useful signal from the received one. The received or obtained data are also called measurements or observations and contain the useful signal and noise. Noise can have natural (environmental, communication channel, electronic equipment, etc.) or artificial (jamming devices, power lines, etc.) sources. Based on the measurements, the receiver determines if an event has occurred. According to [1], the type of detection can be classified into several classes:

- Class I—detection of known signals;
- Class II—detection of signals with unknown parameters;
- Class III—detection of random signals.

The estimation aims to measure the detection's degree of certainty. Some examples of estimation are:

C. Monea and N. Bizon, *Signal Processing and Analysis Techniques for Nuclear Quadrupole Resonance Spectroscopy*, Signals and Communication Technology,
https://doi.org/10.1007/978-3-030-87861-0_2

- Amplitude estimation to identify target's size and the distance to it;
- Spectral estimation to identify the target;
- Frequency estimation to measure the Doppler effect;
- Waveform estimation for signal recovery from noise using optimal filtering.

In its simplest form, the detection problem comes down to determining, based on the received data, whether the events contain the signal of interest or only noise/interference. Since the decision consists in choosing one out of two possible hypotheses, signal with noise or only noise, this is also called the binary hypothesis testing problem. At the receiver level it is desired to maximize the probability of correct detection by using the received data as efficiently as possible. Generally, this becomes a problem of testing multiple hypotheses, and in some applications, it is also called pattern recognition or classification [2].

Two types of trials (experiences, evaluations or measurements) can be considered during an experiment: noise trials, in which only noise is present, and signal plus noise trials (also called signal trials), in which the signal of interest is present, as well as the noise.

The correct identification of the signal following a trial is called a hit. The hit rate is defined as the ratio of the number of correct identifications to the total number of trials:

$$h = \frac{number\ of\ correct\ trials}{total\ number\ of\ signal\ trials} \tag{2.1}$$

In some applications, the hit rate is also called the probability of detection. It is desirable that this ratio be as high as possible. But the hit rate cannot be used to evaluate the performance of the receiver, because it depends on more on aspects of the application than on the detectability itself [3]. In other words, the trial results where only noise is present are neglected. Taking into account both types of trials and results, the following variants are obtained:

Table 2.1 is also called the confusion matrix. The correct identifications are represented by the correctly rejected trials and hits, while the false alarms and miss trials are errors. The evaluation of the receiver is performed based on the hit rate, h, and the false alarm rate, f, defined as follows:

$$f = \frac{number\ of\ false\ alarms}{total\ number\ of\ noise\ trials} \tag{2.2}$$

Table 2.1 Possible detection results

Trial		Response	
		Unidentified signal	Identified signal
	Noise	Correct rejection	False alarm
	Signal	Miss	Hit

The miss rate and the correct rejection rate are determined based on the other two values:

$$missrate = 1 - h \tag{2.3}$$

$$correct\ rejection\ rate = 1 - f \tag{2.4}$$

These types of responses have different names depending on the field or application. For example, the following names are also used:

- hit—true positive;
- false alarm—false positive or type I error;
- correct rejection—true negative;
- miss—false negative or type II error.

The hit and false alarm rates are not sufficient to clearly measure the receiver's detection capability, and a single quantity is preferred to define the detectability of a signal [3]. One way to measure it is to use a model based on the statistical decision theory, which starts from the following assumptions: (1) the evidence regarding the presence of the signal can be described by a single value, (2) the evidence regarding the presence of the signal is subject to random variations and (3) the choice of the answer is made by applying a simple decision criterion based on the amplitude of the evidence. The third hypothesis indicates that the receiver/observer chooses the answer by comparing the evidence of the presence of the signal with a value, called the decision criterion. The decision rule specifies a positive answer, if the evidence is greater than the criterion, respectively a negative answer, if it is less than the criterion, as illustrated in Fig. 2.1.

As can be seen, λ is the decision criterion on the basis of which the response domains are divided. A value greater than λ will generate a positive response, and a lower value will give a negative response.

The evidence's amplitude is a random variable and is denoted X_n, for noise, respectively X_s, for signal. We consider the corresponding probability densities $f_n(x)$ and $f_s(x)$, and the distribution functions $F_n(x)$ and $F_s(x)$. As seen in Fig. 2.1, the probability of false alarms and hits are determined by finding the area under the probability densities curves for values greater than λ. Expressed mathematically, the two probabilities are:

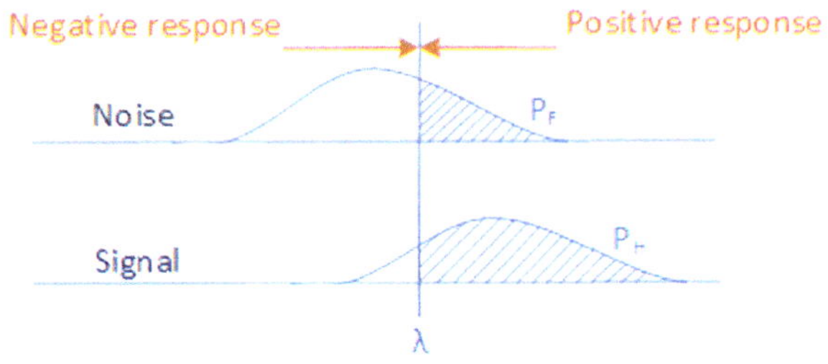

Fig. 2.1 Probability distributions of noise and signal plus noise, indicating the decision criterion, as well as the false alarm and hit probabilities

$$P_H = P(X_n > \lambda) = \int_{\lambda}^{\infty} f_s(x)dx = 1 - F_s(\lambda) \quad (2.5)$$

$$P_H = P(X_s > \lambda) = \int_{\lambda}^{\infty} f_s(x)dx = 1 - F_s(\lambda) \quad (2.6)$$

The choice of the criterion (λ) influences the hit and false alarm probabilities. If it is too low, then both rates are high. Otherwise, if it has a higher value, then the rates will be low. The signal's detectability is given by the placement of the two distributions and the degree of their overlap, while the receiver's/observer's interpretation strategy is given by the criterion. By mathematically expressing the distributions of the two random variables, predictions of the hit and false alarm probabilities can be made.

By graphically displaying the success rate vs. the false alarm rate for different values of the decision criterion, the *Receiver Operating Characteristic* (ROC) is obtained. The ROC curve was originally in the 1940s in order to evaluate the radar operator's capacity to correctly identify enemy aircraft. Today they are generally used to evaluate classification models in many fields, such as psychology, electronics, medicine, machine learning, and so on. An example of a ROC curve is illustrated in Fig. 2.2.

When the value of the criterion λ is small, the probabilities P_H and P_F are high, giving a point closer to (1, 1). This is not desirable, as the receiver will indicate false alarms with high probability, increasing the reception error. When λ increases, the two probabilities decrease, approaching the point (0, 0). This situation is also not desirable, because it decreases the hit probability. The aim is to identify the value of the decision criterion that determines a point (P_F, P_H) as close as possible to the point (0, 1). By increasing, e.g., the signal's amplitude, a family of curves is obtained, which are closer to the point (0, 1), indicating better receiver performances in conditions where the signal is strong enough to allow error-free detection.

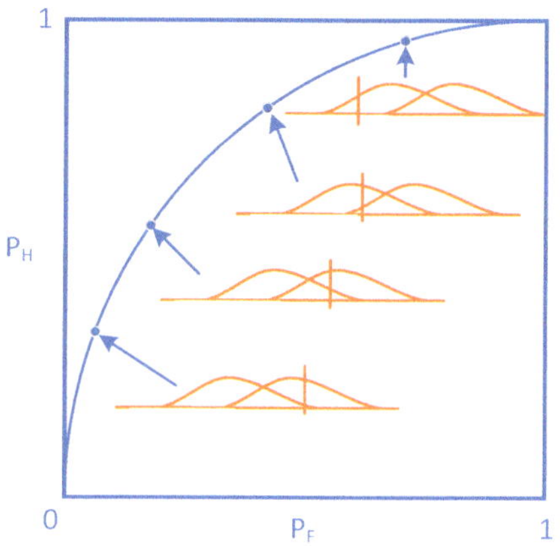

Fig. 2.2 Receiver operating characteristic

Signal processing and estimation methods aim to differentiate information-bearing signals from random ones and to quantitatively evaluate them, in order to increase the probability of detection. The processing of the signal affected by noise/interference involves its processing so that the effect of the noise is reduced enough to allow the signal to be identified with a particular degree of certainty. The noise sources can be: the transmission medium/communication channel, electrical/electronic equipment, thermal noise, etc. Generally, one way to increase the noise immunity of an electronic system is to increase the signal and processing methods' complexity.

In practice, the processing is often performed on small amplitude signals, in conditions where the signal-to-noise ratio is much lower than 1; in this case, the main problem is represented by the optimal reconstruction of the signal at the receiver [4]. In these cases, the processing methods aim to detect the signal "buried" in noise and depend on the amount of received data; also, a higher signal-to-noise ratio value is required for its estimation [2]. Noise, interference and signal distortion result in decreased detection probability. The receiver must be designed to make the most efficient use of previously (a priori) known information regarding the noise, signal, channel or interference. Among the most common signal processing and analysis methods applied in the detection theory, one can mention: power spectrum analysis (periodogram, fast Fourier transform, estimation methods—maximum entropy, moving average, ARMA, high resolution methods—Pisarenko, MUSIC, SPIRIT), filtering (Wiener, Kalman, RLS, LMS, matched filter, etc.), statistical algorithms for detection (criteria for simple hypotheses—Neyman-Pearson, Bayes, maximum likelihood, minimum probability of error, maximum a posteriori or minmax, respectively multiple hypotheses).

There are many applications for signal detection, and the methods and equipment used are diverse. These include: (1) metal detection, (2) georadar to locate, identify, and visualize variations of the soil's electrical and magnetic properties, (3) microwave tomography to monitor hidden structures beneath a surface and to detect hidden objects, (4) detection of vibrations in the ground in order to analyze its structure, underground objects or to detect the presence of human or animal activity, (5) remote multispectral detection to perform geological and geographical measurements, etc.

2.2 Techniques for Detecting Prohibited Substances

The detection of prohibited substances is an application of the theory of detection and plays an important place in the field of security. The detection of explosives is of greater importance due to material damage and casualties that they can cause either accidentally or when used for terrorist purposes. For this reason, the following text will focus on the description of the techniques used to detect explosives, mentioning, where appropriate, the applicability for identifying other types of prohibited substances.

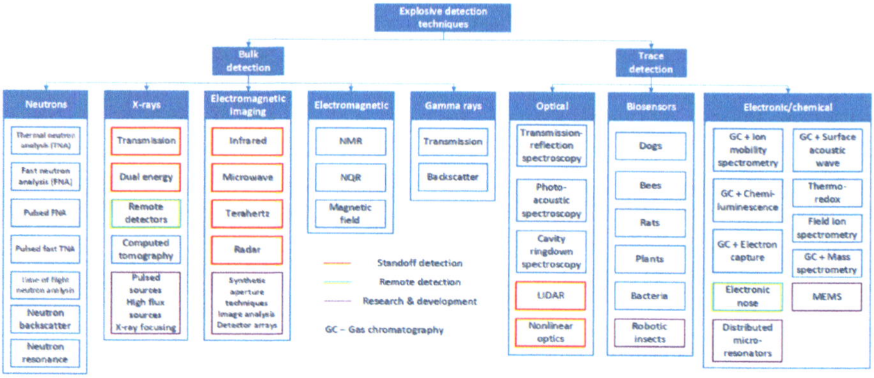

Fig. 2.3 Techniques for bulk and trace detection (NMR—nuclear magnetic resonance, NQR—quadrupolar nuclear resonance, Electromag.—electromagnetic, Spectr.—spectrometry, CG—gas chromatography, MEMS—microelectromechanical systems) [5] (© 2018 J. of Electrical Eng., Electronics, Control and Computer Science)

The explosive detection techniques are categorized according to the substance sought, as follows: mass (bulk) detection, respectively trace detection techniques [5]. The first identify a mass of material with certain properties that identify it as explosive, such as, e.g., high density of content of nitrogen [6]. The trace detection techniques detect the presence of an explosive substance by collecting and analyzing small amounts of vapor or particles [7]. In this case, the general approach is to search for a compound that exists in a target library, rather than a property of the material. Generally, the trace detection techniques give lower false alarms; however, they do not allow the identification of the substance's location in the scanned item.

Figure 2.3 shows the classification of the explosive detection techniques. The existing techniques for the detection of prohibited substances. The following color code is used: red—techniques that allow standoff detection, green—techniques for remote detection, purple—techniques under research and development.

2.2.1 *Trace Detection Techniques*

The ion mobility spectrometry (IMS) method measures the sample's ionized vapors speed when they move towards the actual detector, in an electric field [8]. The working principle involves collecting particles either by wiping the sample with another surface or by collecting the air around it. The particles are ionized in a dedicated zone located inside the detector and the newly formed negative ions inside an electrical field (drift zone). The ions' drift time depends on their electric charge, mass and size, and lasts in the order of milliseconds. The substance can be precisely identified by measuring the time the ions require to traverse the drift zone, which depends on their size, size and electric charge.

IMS detectors are manufactured as handheld or office equipment, as well as portals. Their advantages are given by their portability, low detection limit (in the order of picograms) and the possibility to scan people [7]. However, they have a long scan time, taking several samples/min. [9], they contain an ionization source, which may be harmful for the operators, and may underperform in the case of substance combinations. One solution used to avoid this problem is to use gas chromatography to separate the molecules before they are analyzed by the IMS technique. IMS detectors can identify explosives and drugs, such as IONSCAN 500DT, produced by Smiths Detection, which can identify military and commercial explosives (EGDN, HMTD, HMX, NG, PETN, RDX, TATP, tetryl, TNT) and narcotics (amphetamine, cocaine, heroin, ketamine, MDMA, methamphetamine, THC).

Chemiluminescence is the generation of electromagnetic radiation (ultraviolet, visible or infrared) under the conditions of a chemical reaction [8]. Usually, the infrared radiation generated by NO_2 or NO_3 groups is detected, because these are the most common among explosives. An example is the E3500 chemiluminescence detector, produced by Scintrex Trace Corporation. Its list of detected substances contains a wide range of explosives, such as RDX, TNT, PETN, dynamite, Semtex, black powder, C-4, urea nitrate, ammonium nitrate, NG, EGDN, DMNB, TATP, chlorates, peroxides and others.

These detectors are similar to IMS, and their advantages are given by their portability and the fact that they do not use (radioactive) ionizing sources. Their major disadvantage is that they are limited to substances that contain nitrogen groups, and may give false alarms when scanning, e.g., fertilizers. Usually, they are complemented with a gas chromatograph.

The thermo-redox method draws air into the detector and pyrolyzes the traces in the air. No carrier gas is required and the detectors are handheld. They also require minimal operator training, and their cost and consumables are relatively low. Thermo-redox suffers from the fact that it cannot identify the explosive's type and gives false alarms in case of, e.g., fertilizers [9]. An example of a commercial detector is EVD-3000+ , produced by Scintrex Trace Corporation, which has a detection time of maximum 10 s and can identify RDX, PETN, C-4 etc.

The surface acoustic wave method is based on depositing the substance's particles on the detector and measuring the frequency differences caused by them. These detectors can be used for substances that are not explosives, are safe for the operators because they do not integrate any ionizing source, are small (portable devices) and have a moderate cost. Their main disadvantages are their non-specificity and the fact that they require a carrier gas cylinder [9]. The zNose Model 4200, produced by Electronic Sensor Technology, is an equipment based on the surface acoustic wave method which requires 30–60 s per scan.

The mass spectrometry technique identifies the substances based on the masses of the molecules and atoms. Their working principle is based on measuring the mass-to-charge ratio of the ions [8]. The mass spectrometer may allow the unique identification of substances, including complex combinations, when supplemented with a gas chromatograph [7]. Mass spectrometers are highly specific and are especially used in the laboratory. They also exist in the form of portable devices. Their disadvantages are the facts that that they have a high cost, require trained operators, require a long time for initialization and scanning. One example is the highly integrated 4500 MiD spectrometer, manufactured by Microsaic Systems, which requires no external devices or components.

Dogs are probably the most used method of biological detection of prohibited substances, and are considered to be some of the most sensitive detection "tools" [6]. Generally, dogs focus on either the explosive, a contaminant or a component of the explosive. Dogs are highly mobile, can identify the trace's source with very good sensitivity and specificity. According to [10], other biological methods include bees, mice, ants, bacteria and plants.

2.2.2 *Bulk Detection Techniques*

The X-ray techniques are widely used for luggage or containers screening. They detect the materials that have explosive-like characteristics, such as high density and low atomic number.

In principle, X-ray devices contain a radiation source and a detector. The source is responsible to illuminating the scanned item or person. The transmitted X-rays through the target are measured when the source and detector are placed in opposite positions. Otherwise, the reflected X-rays by the target are measured.

The advantages of X-ray scanners are the discrimination of materials based on their shape, metal detection and relative low cost. Their disadvantages are the use of an ionizing source, respectively the difficulty of separating objects from the image. There are many X-ray equipment, including those produced by Smiths Detection, such as HI-SCAN 6040C and HI-SCAN 6040 CTiX, for scanning luggage and packages.

Millimeter wave scanning is a technique that uses pairs of transceivers which rotated around the scanned person. An image of the person is created from the radiation reflected by its skin and the objects concealed under its clothing. The working frequency range is 30–300 GHz. Millimeter wave scanning uses non-ionizing radiation. An example is the ProVision 2 scanner manufactured by L3 Security & Detection Systems.

Neutron-based nuclear techniques are generally based on the characteristic emission from which the object which is investigated. The target is illuminated by a low-energy neutron flux which is absorbed by its atomic nuclei, which emit high-energy gamma rays. The energy us in the range of 10.8 meV. Their detection indicates the presence of substances that contain nitrogen, including explosives, narcotics and toxic substances. This method is used for luggage and cargo screening, since gamma rays can pass through many objects (including metallic objects). However, this scanning method cannot be used on humans, as exposure to neutrons affects human health.

2.3 Conclusions

This chapter aimed to be a bibliographic study of methods and equipment for the acquisition and analysis of signals for the identification of dangerous and contraband substances. An introduction was made in the theory of signal detection and estimation, as a general framework, the presentation gradually focusing on the main detection techniques.

Signal detection and estimation theory has a wide range of applications, including: radar, sonar, communications, seismic detection, model recognition, metal detection, remote measurement, landmine detection, geophysical explorations (prospecting), prediction, etc. The signal processing techniques used in this field are also varied: signal detection and estimation algorithms, imaging processing solutions or machine learning-based signal analysis. One application is the detection of prohibited substances such as explosives, drugs and toxic substances, which is also an important research direction in the field of security.

The main techniques used in commercial and laboratory equipment for the detection of prohibited substances, including those for trace and bulk detection, were presented. The study aimed to describe the physical principles on which they are based and to comparatively present their advantages and disadvantages.

References

1. Hippenstiel, R.D.: Detection Theory: Applications and Digital Signal Processing. CRC Press (2002)
2. Kay, S.M.: Fundamentals of Statistical Signal Processing, vol. II. Pearson (1998)
3. Wickens, T.D.: Elementary Signal Detection Theory. Oxford University Press (2001)
4. Tuzlukov, V.P.: Signal Detection Theory. Springer Science+Business Media (2001)
5. Monea, C.: Techniques and equipment for detection of prohibited substances: a brief overview. J. Electri. Eng. Electron. Control Comput.er Sci. **4**(4), 7–16 (2018)
6. Marshall, M., Oxley, J.: Aspects of Explosives Detection, First edn. Elsevier Science (2008)
7. National Research Council of The National Academies—Committee on the Review of Existing and Potential Standoff Explosives.: Detection Techniques, Existing and Potential Standoff Explosives Detection Techniques. The National Academies Press (2004)
8. Yinon, Y.: Counterterrorist Detection Techniques of Explosives. Elsevier (2007)
9. Theisen, L., Hannum, D.W., Murray, D.W., Parmeter, J.E.: Survey of Commercially Available Explosives Detection Technologies and Equipment (2004)
10. Cardona, L., Jiménez, J., Vanegas, N.: Landmine detection technologies to face the demining problem in Antioquia. Dyna. **81**(183), 115–125 (2014)

Chapter 3
Nuclear Quadrupole Resonance Spectroscopy

3.1 The Principle of Nuclear Quadrupole Resonance

Nuclear quadrupole resonance, abbreviated NQR, is a solid spectroscopy technique used in chemical analysis. It was first observed in 1949 by Dehmelt and Krüger [1]. The NQR phenomenon occurs in solid substances (powders or crystalline substances) composed of atoms with quadrupolar nuclei, having a quantum spin number (I) greater than 1. NQR does not require a static external magnetic field, as opposed to nuclear magnetic resonance (NMR).

Nuclear spin represents an intrinsic form of angular moment associated with the rotational motion of a particle around its own axis. It is a quantum property and can only take integer or semi-integer values. The spin quantum number represents the set of spins of the particles that make up the atomic nucleus (protons and neutrons) and determines the number of quantum states in which the nuclei of the substance can be found.

Quadrupolar nuclei have a nuclear quadrupole electric moment (Q), which characterizes the deviation of the nucleus' positive electric charge distribution from the spherical distribution. If the electric charge distribution has an elongated shape, then Q is positive. Conversely, if it has a flattened shape, then Q is negative. The nuclear quadrupole electric moment, resulting from the non-spherical distribution of the positive charges of the nucleus, interacts with the electric field gradient created by the electric charges surrounding the nucleus (mainly the negative charges of electrons and, to a lesser extent, the positive charges of neighboring nuclei). These interactions split the energy levels where transitions can take place. The lowest energy level corresponds to the orientation in which the highest amount of positive nuclear charge is closest to the highest negative charge density in the vicinity of the nucleus.

C. Monea and N. Bizon, *Signal Processing and Analysis Techniques for Nuclear Quadrupole Resonance Spectroscopy*, Signals and Communication Technology,
https://doi.org/10.1007/978-3-030-87861-0_3

The transitions are achieved by applying an electromagnetic field with the frequency:

$$\nu = \frac{\Delta W}{h} \tag{3.1}$$

where h represents Planck's constant and ΔW is the difference between the energy levels. This selective absorption of electromagnetic radiation is called nuclear quadrupolar resonance.

The energy levels between which transitions can take place and, implicitly, the transition frequencies, depend on the field gradient q due to the distribution of electrons in the molecule and the nucleus' electric quadrupole moment. These levels are expressed mathematically as follows for an unitary spin number:

$$E_z = \frac{-2eQq}{4I(2I-1)},\ E_x = \frac{eQq(1-\eta)}{4I(2I-1)},\ E_y = \frac{eQq(1+\eta)}{4I(2I-1)} \tag{3.2}$$

where η is called the asymmetry parameter, q represents the EFG's component along the z-axis, and e is the elementary electric charge. By convention, the z-axis is considered the direction with the largest component of the EFG. The transition frequencies are expressed as follows:

$$\nu_+ = \frac{eQq(3+\eta)}{4h},\ \nu_- = \frac{eQq(3-\eta)}{4h},\ \nu_0 = \frac{eQq\eta}{2h} \tag{3.3}$$

The transition frequencies between the different energy levels are illustrated in Fig. 3.1.

In case (a), η is non-zero, resulting in different transition frequencies. When the asymmetry parameter is zero (b), only one frequency $\nu_+ = \nu_-$ is obtained. If it is unitary (c), then two frequencies are obtained, ν_+ and $\nu_- = \nu_0$.

The electric quadrupolar moment represents a combination of the charges in the nucleus and is specific and constant for a certain isotope. NQR resonance frequencies depend on the gradient of the electric field in the vicinity of the nuclei. Also, the frequency band in which the resonant frequencies are located is determined by the nature of the nuclei with non-zero quadrupolar moment. Therefore, the NQR response signal is specific to each substance, and its characteristics are strongly dependent on the nature of the atoms and the molecular structure (the type of chemical bonds that exist in the molecule). In other words, this technique allows the detection of a

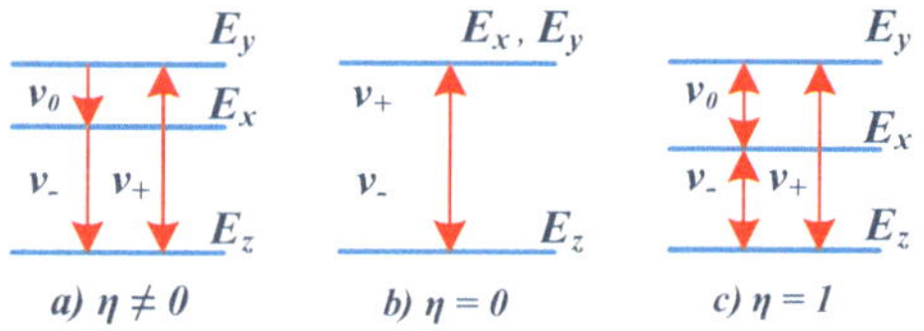

Fig. 3.1 NQR transition frequencies in the cases: **a** $\eta \neq 0$, **b** $\eta = 0$ and **c** $\eta = 1$

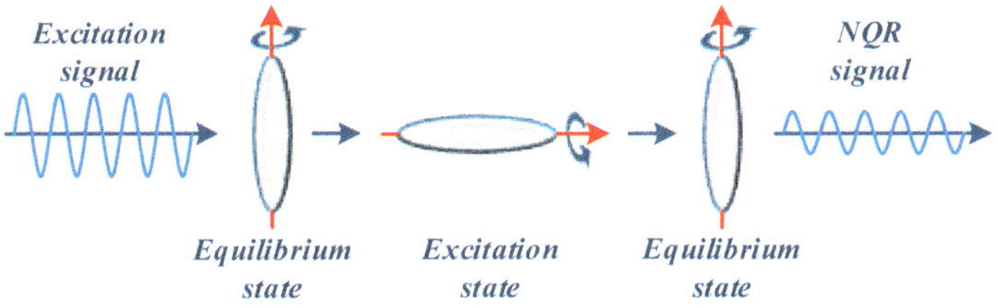

Fig. 3.2 NQR excitation process

quadrupolar nucleus of a molecule in a given chemical state. Substances that do not contain nuclei with a non-zero quadrupole moment exhibit no NQR effect. Depending on the value of η, a substance can have 1 to 3 resonant (transition) frequencies.

In liquids and gases, the average of the quadrupole interaction in relation to the molecular motion is very small, leading to a very weak effective interaction. In other words, the molecular motion makes the time average of these interactions zero. Therefore, nuclear quadrupole resonance cannot be observed in liquids and gases [2].

The excitation process is illustrated in Fig. 3.2.

At thermodynamic equilibrium, the energy levels are approximately equally populated. Considering the ^{14}N isotope at room temperature, the population difference reaches 1 nucleus to 1 million nuclei, in the most favorable case. The excitation of the substance results in larger differences between the populations for each energy level, which will induce transitions between the levels. When the atoms return to equilibrium, after the excitation has ended, the excess energy released will take the form of a measurable radio frequency signal, called *free induction decay* (FID), illustrated in Fig. 3.3.

The FID signal is generated due to the precession of the nuclear spins. It has an exponentially damped shape, and the frequency is equal to that of the excitation signal. The signal decay rate is given by the transverse or spin–spin relaxation time (T_2); in this case, the relaxation means the return of the excited nuclei to the steady state by the by loss of coherence due to spin–spin interactions or thermal scattering. For powders, its effective value is denoted T_2^*. This constant can take values in the range of μs to ms.

The signal generated by a nucleus is too small to be detected, but the sum of the signals from all the excited nuclei produces a signal large enough to be observed.

Fig. 3.3 Free induction decay signal

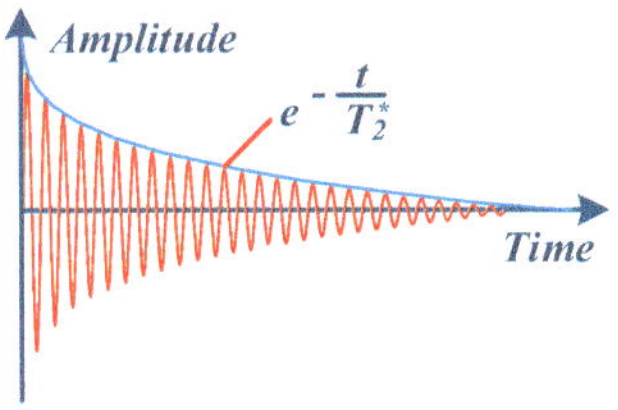

There are variations of the electric field gradients within the substance, due to the inhomogeneities introduced by impurities, the imperfect crystalline structure, the thermal gradients or, in the case of powders, the existence of crystals with different orientations. As a result, certain nuclei may receive different electric fields and, consequently, may emit signals with slightly different frequencies. This leads to the loss of phase synchronization of the signals emitted by the quadrupolar nuclei, resulting in the exponentially damped form of the FID signal (signal decay).

The relaxation time required for the excited nuclei is denoted T_1. This constant is called the longitudinal relaxation time or spin–lattice relaxation time, and has values in the range of ms to tens of s. It represents the time required for the excited nuclei to dissipate their energy by interacting with the crystal lattice. The return to thermal equilibrium involves the release of excess energy to neighboring nuclei and molecules, through collisions, rotations or electromagnetic interactions. T_1 determines the rate at which the energy absorbed by the substance can be dissipated as heat and thus the rate at which the excitation can be repeated.

In the case of pulse excitation, in order to obtain a stronger response, the excitation is performed with several successive pulses of different durations and the reception aims to acquire the generated echo signals. In this case, the parameter T_2 represents the echo train decay time, and its effective value, T_{2e} has values in the range of tens of ms to a few s, and is greater than T_2^*.

The NQR resonance frequencies have different values associated with these time parameters and are characteristic to each substance. These parameters depend on temperature and may vary nonlinearly in some cases.

Taking into account the NQR principle, several factors affecting signal detection can be mentioned.

- **Detection distance**

The distance affects the SNR, because the signal strength decreases as the distance between the receiver and the substance increases.

- **Excitation frequency**

There is a proportional relationship between the power of the response signal and its frequency; substances with resonant frequencies below 1 MHz respond weaker than others. According to [3], the SNR increases by $\nu^{3/2}$.

- **Receiver noise**

The thermal or inherent noise [4] at the receiver, including the detection coil and preamplifier, limits the SNR [5].

- **Substance temperature**

The temperature affects the molecular motion, and, consequently, the EFG in the nuclei and the resonant frequencies [6]. The variation with the temperature introduces uncertainties regarding the appropriate excitation parameters, and the excitation with frequencies that do not coincide with those associated with the substance's temperature will generate a weak response. For acceptable results, it is necessary to determine the temperature of the substance with an accuracy of ± 5 K [7].

- **Substance nature**

Powders contain crystals with different orientations, leading to excitation differences between them. This reduces the signal strength by 57% compared to a single crystal [5]. Also, polymorphs, such as TNT, have different resonance frequencies depending on the crystallization form and it is possible that only a part of their nuclei are excited when scanning is performed on a specific frequency [8].

- **Substance quantity**

In large quantities, the excited substances will have more nuclei that will generate a response signal, and overall, its amplitude will increase (implicitly, the SNR).

- **Excitation sequence**

The parameters of a pulse excitation sequence have to be adjusted in order to maximize the response signal's amplitude (sequence type, pulse duration, pulse duration, sequence duration, frequency, excitation power). Generally, the amplitude of the response increases with the excitation power.

- **RF interferences**

One of the characteristics of the NQR technique is the fact that the response signal of the substance is very weak, having the amplitude in the order of tens of nV, comparable to the receiver's thermal noise level (very low SNR). Thus, it can be easily affected by radio frequency interference (RFI) or noise signals such as piezoelectric and magneto-acoustic responses (coherent noise [4]).

RFI is classified in far and near field interference and is typically caused by shortwave equipment, the detector itself or by sample impurities [9]. RFI reduction solutions include shielding, the use of the gradiometric coil, the use of matched filtering and innovative signal processing algorithms. The last two methods are mainly used in applications such as landmine detection.

- **Detection circuit quality factor**

The NQR response signal has amplitudes in the order of nV [10], and its detection requires the use of a circuit with a high-quality factor (Q). On the other hand, a high Q factor increases the duration of the transient signals (for example, the coil ringing, which represents the oscillations in the detection coil due to the inverse electromotive voltage that appears when an excitation pulse ends) [11]. Its duration can reach hundreds of μs when working at low frequencies (below 2 MHz), making it difficult to detect substances with short T_2^* relaxation times [12].

There are many isotopes found in nature that can be analyzed using NQR: ^{14}N, ^{35}Cl, ^{63}Cu, ^{65}Cu, ^{17}O, ^{59}Co, ^{47}Ti, ^{49}Ti, ^{55}Mn, ^{27}Al, ^{115}I, ^{71}Ga etc. ^{14}N is found in most dangerous and contraband substances, and NQR is used to radio "fingerprint" them. More than 10,000 substances have been studied so far and no two substances have been identified with the same resonance frequency [13].

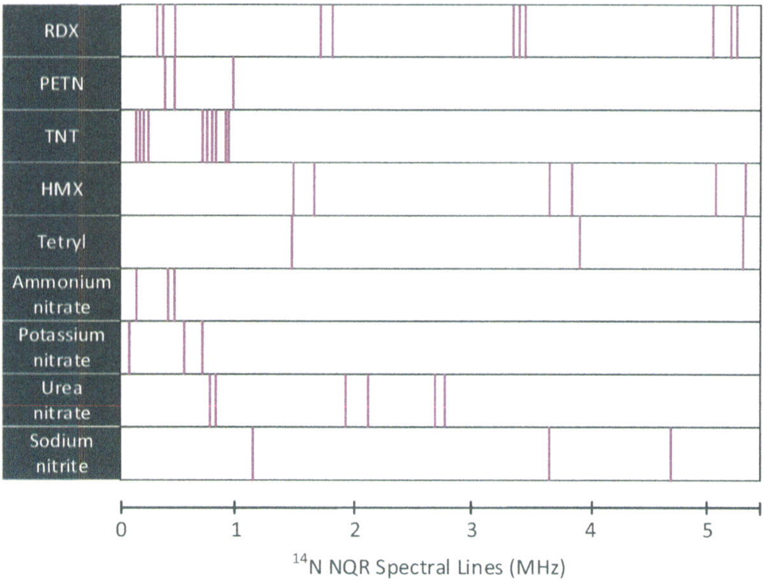

Fig. 3.4 NQR resonance frequencies for different substances

The NQR resonance frequencies depend on temperature, vary differently for each substance, and, in some cases, nonlinearly. NQR data for different substances is presented in Appendix 1 and the drift of the time parameters and resonance frequencies with the temperature can be observed.

Figure 3.4 illustrates the spectral lines for several substances that contain the isotope ^{14} N.

As seen, certain substances, such as RDX, TNT and HMX, have several spectral lines. They are also due to the fact that the substances differ from one manufacturer to another or may have different crystalline structures (for example, TNT may be monoclinic or orthorhombic). Also, the longitudinal relaxation time shows a nonlinear variation with temperature [3].

NQR is known for its high specificity. Therefore, it is used for chemical analysis. Also, it is applied for luggage scanning and landmine detection, as well as pharmaceuticals analysis [14], mineral analysis and temperature measurement [15].

3.2 Signal Acquisition Methods Applied in Nuclear Quadrupole Resonance Spectroscopy

The NQR signal acquisition (detection) methods refer to the excitation method applied for obtaining the substance response and the way its reception is performed,

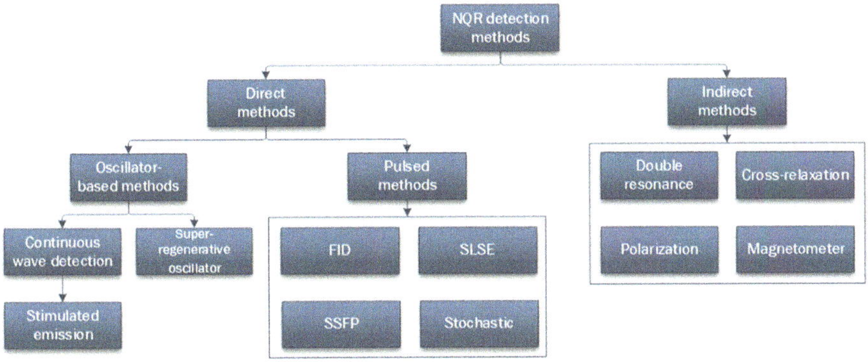

Fig. 3.5 NQR detection methods (FID—free induction decay, SLSE—spin-locked spin echo, SSFP—steady-state free precession) [16] (© 2019 IEEE. Reprinted, with permission, from C. Monea, The use of nuclear quadrupole resonance spectroscopy for detection of prohibited substances: Techniques and equipment, 11th Int. Conf. on Electronics, Computers and Artificial Intelligence, pp. 1–6, 2019)

respectively. As illustrated in Fig. 2.10, they fall in two broad categories according to the intensity of the investigated substances' response [16]:

- Direct methods—applied for substances with strong response;
- Indirect methods—applied for substances with low resonance frequencies (less than 1 MHz) (Fig. 3.5).

Continuous wave (CW) detection uses a Robinson or marginal oscillator. When the oscillator's frequency is equal to the substance's resonance frequency (with the substances placed inside or near the resonance circuit), the amplitude of the oscillations will decay. The super-regenerative oscillator (SRO) was the method applied to demonstrate the NQR phenomenon. It is based on reaching a pulse saturation voltage when the substance's response is emitted between the pulses; thus, the increase in pulse duration will indicate the detection of the response signal.

Continuous wave excitation can be performed by using stimulated emission, as described in multiple [17–19] and papers [20, 21]. It is based on the Rabi formulation and assumes that the substance will periodically absorb and emit radiation, at time intervals T_1 [17, 18]. The nuclei's transitions are achieved by stimulated emission. Low-power excitation is possible and the receivers typically uses matched filters [19].

The pulse-based methods are widely used since they increase the signal-to-noise ratio and reduce the time required for signal acquisition. The multi-pulse sequences are classified into spin-locked spin echo sequences (SLSE), also called spin-locked multi-pulse (SLMP), and steady-state free precession (SSFP), respectively. Generally, these types of sequences increase the signal-to-noise ratio with the square root of the number of scans, decrease the scanning time and enables the measurement of the relaxation times. The optimal parameters of a sequence differ from one substance

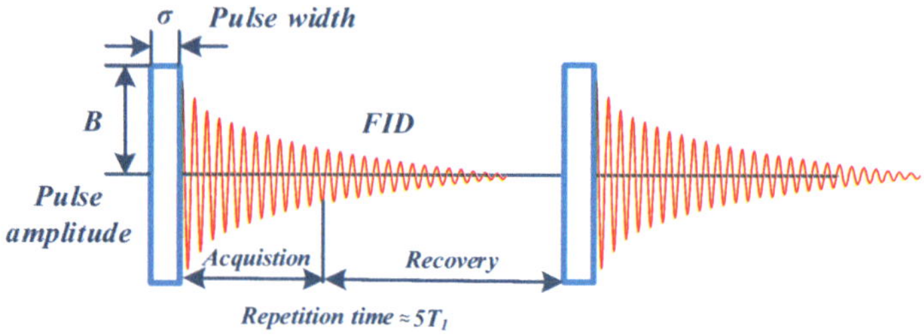

Fig. 3.6 FID sequence [16] (© 2019 IEEE. Reprinted, with permission, from C. Monea, The use of nuclear quadrupole resonance spectroscopy for detection of prohibited substances: Techniques and equipment, 11th Int. Conf. on Electronics, Computers and Artificial Intelligence, pp. 1–6, 2019)

to another, depending on the detection conditions. Also, the optimal sequence is not identical for multiple substances. The following describes the types of excitation sequences used in the NQR spectroscopy and exemplifies some of them.

The **FID sequence** is composed of a transmitted pulse followed by the response reception. A waiting time of $5T_1$ is required to allow complete relaxation of the substance. This sequence has the drawback of a long scanning time; thus, it is used for substances with small T_1. The angle of rotation of the nuclear spins has the following expression [7]:

$$\theta = \sigma \gamma B \frac{180}{\pi} \tag{3.4}$$

where σ is the pulse duration (width), in s, B is the magnetic flux density near the nucleus, in T, and γ is the gyromagnetic ratio, expressed in $rad/(sT)$. Figure 3.6 shows the FID sequence.

The pulse duration must be chosen carefully, because the intensity of the response is proportional to it. In the case of monocrystalline substances, the maximum response is obtained for $\theta = 90^{\circ}$, if the magnetic field of the excitation pulse is aligned with the nuclear spin. In the case of powders, where there are many crystals with different orientations, the maximum signal is obtained for $\theta = 119^{\circ}$, also called "equivalent 90° pulse" [22]. Although it can be calculated, in practice, the optimal duration is determined experimentally. Also, another consequence is that the pulse width must be adjusted when the distance between the substance and the detection coil changes, because the magnetic flux density decreases as the distance increases. In other words, as the distance increases, the duration of the pulse must be increased. In addition, if the excitation power increases, the pulse duration decreases.

The **SLSE sequences** aim to generate and receive echoes. While FID is the damped signal that occurs immediately after the excitation pulse stops, the echo is a regenerated signal with the maximum intensity between the excitation pulses. The echo (also called spin echo—SE, or Hahn echo) occurs when a pulse with double duration is applied before the substance relaxes completely (returns to equilibrium). If the longitudinal relaxation did not eliminate the signal, additional echoes can be generated, as shown in Fig. 3.7.

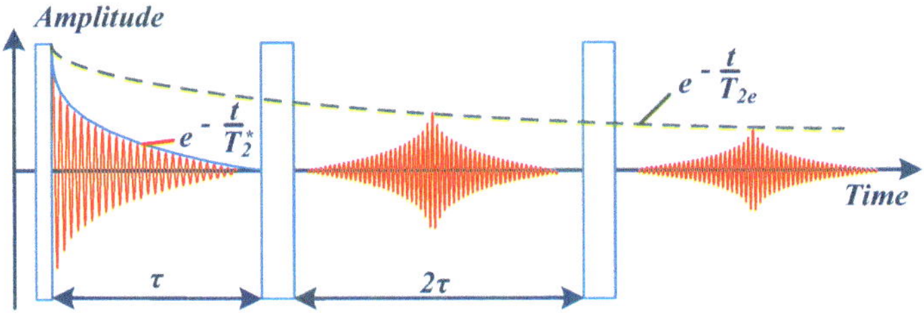

Fig. 3.7 SLSE sequence [16] (© 2019 IEEE. Reprinted, with permission, from C. Monea, The use of nuclear quadrupole resonance spectroscopy for detection of prohibited substances: Techniques and equipment, 11th Int. Conf. on Electronics, Computers and Artificial Intelligence, pp. 1–6, 2019)

The amplitude of the echoes has an exponential decay with the (effective) rate T_{2e}; this is higher than that of the FID signal, T_2^*. T_{2e} limits the length of the spin-echo sequence. SLSE sequences are noted as follows:

$$\theta_{\varphi_1} - \left(\tau - \theta_{\varphi_2} - \tau\right)_N \tag{3.5}$$

where N is the number of pulses, 2τ is the repetition time, φ_1 is the first pulse's phase (also called preparation or $\pi/2$ pulse, φ_2 is the refocusing pulse's phase, and θ is the rotation angle (correlated with the pulse duration). The duration of the first pulse must correspond to a rotation angle of 90°, but the following pulses may have different durations.

Phase-rotating (cycling) pulse sequences have been proposed to reduce unwanted signals like piezoelectric noise [3]. These sequences are adequate when T_1 is greater than τ and much higher than T_2. In the literature, they are also called SE, pulsed spin locking (PSL), Carr-Purcell-Meiboom-Gill (CPMG) sequences, when applied in NMR spectroscopy, or Carr-Purcell-Meiboom-Gill Quadrupolar (QCMPG), when applied only to quadrupole nuclei (in NQR spectroscopy).

In practice, the temperature of the substance may be unknown. The resonance offset sequences have been proposed to identify the response frequency deviation from the excitation frequency [10]. In other words, the difference between the excitation frequency used and that given by the temperature of the substance is observed. Thus, the SLSE sequence becomes phase-alternated spin-locked spin echo (PASLSE). Another particular case is the WURST-QCMPG sequence, which can be used for substances with unknown temperature [23]. The difference from other sequences is that the frequency is continuously swept in a band in the order of kHz around the center (resonance) frequency during the excitation pulse; this enables the acquisition of the response in the same time as an SLSE sequence, if its temperature differs by several degrees compared to the temperature corresponding to the center frequency.

SSFP sequences, unlike SLSE, use pulses with equal durations and intervals. SSFP sequences do not allow the substance to completely return to the steady state, and maintain it in a state called steady-state free precession. The response takes the form of a variable-amplitude continuous signal. Although the signal between two pulses is complex, it may resemble the merge between the end of the FID signal and the beginning of an echo. The SSFP sequences are noted as follows:

$$\left(\tau - \theta_{\varphi} - \tau\right)_N \tag{3.6}$$

where N is the number of pulses, 2τ is the repetition time, φ is the pulse phase, and θ is the rotation angle (correlated with the pulse duration).

These sequences are used when T_1 is low and $T_1 \cong T_2$ [3]. In the literature, they are also called phase-alternated pulse sequence (PAPS) sequences, when $\varphi = 180^{\circ}$, and non-phase-alternated pulse sequence (NPAPS) or strong off-resonance comb (SORC), when $\varphi = 0^{\circ}$, respectively. In the latter case, an excitation frequency different from that of the substance is used.

The coil ringing can be canceled by combining the echoes from four SSFP sequences: NPAPS, PAPS, PAPS and NPAPS, with the excitation pulse phases shifted by π with respect to the previous sequences [10]. In [24, 25] modified SSFP sequences are proposed for the elimination of coherent magneto-acoustic, piezoelectric noises and coil ringing. They are based on the phase rotation technique and the addition of additional pulse blocks.

Stochastic excitation uses short RF pulses having pseudo-/random phases, which keep the substance in an excitation state [26]. At reception, the acquired signal is cross-correlated with the initial sequence and the FID signal is obtained by applying the Hadamard transform. Stochastic excitation allows excitation at low powers. In [26], a pulse width of 5 μs was applied, with the time between pulses of 40 μs.

The weakly responsive substances can be acquired using indirect methods. Double resonance consists in observing the weak response signal emitted by one type of nuclei, while the strong signal emitted by the second type is detected. The cross-relaxation method is similar to the previous one, but does not perform the excitation of the quadrupolar nuclei. Also called cross-polarization, the polarization method increases the signal-to-noise ration by applying a static magnetic field before performing the actual NQR excitation. Its advantage is the reduction of the relaxation time, allowing a larger number of scans to be performed in the same period of time. The magnetometer is another method that allows the detection of very weak NQR response signals. One of the magnetometers used is superconducting quantum interference device (SQUID). It can be considered a transducer that performs the conversion of the magnetic flux into electrical voltage, with very low thermal noise. Its advantage is that it can be used in any type of NQR measurement, either continuous wave or pulse-based, and allows the detection of broadband signals, because it does not require working at the resonance frequency. This magnetometer has the disadvantage of high cost associated to the cryogenic cooling equipment.

3.3 Equipment Used in Nuclear Quadrupole Resonance Spectroscopy

The equipment used in NQR spectroscopy varies from laboratory equipment, detectors or prototypes, to commercial equipment.

One can mention the excitation and acquisition systems Spincore [27] and Redstone [28], produced by the American companies SpinCore Technologies and Tecmag, respectively. Tomco Technologies is a manufacturer from Australia focused on pulsed or continuous wave radio amplifiers for special applications, including NQR. For NQR, the product range includes amplifiers in the range 250–4 kW, and the frequency band from 0.1–810 MHz; an example is the BT00250-AlphaS amplifier, with a peak power of 250 W and the working frequency band 100 kHz–30 MHz [29]. Also, it is worth mentioning the company L3 Narda-MITEQ, which produces multiple radio frequency equipment, such as PIN diode switches, amplifiers, multipliers, etc.

At the laboratory level, several portable equipment have been developed for landmine detection [10, 30], pharmaceuticals validation [14, 31] and explosives identification [32, 33]. E.g., the detector presented in [7] is portable and light. It integrates an FPGA circuit for generating the excitation sequences; it also processes the received signal, as well as a bandpass filter and reception amplifier. The coil has a gradiometric shape to reduce RFI.

CONPHIRMER was a European project that aimed to develop a detector for drugs based on NQR. The prototype is designed to detect triazole, paracetamol and paradichlorobenzene (used as a substitute for naphthalene). It uses two distinct and interchangeable coils, for low and high frequencies, respectively, and has a peak power of 200 W.

The use of indirect techniques has been investigated to improve detection, such as polarization [34, 35], optically pumped magnetometer [36], atomic magnetometer [37, 38], quantum interference superconducting device [39, 40] and giant magnetoresistive sensor [41]. Special solutions to increase robustness to noise [42] and broadband scanning [43, 44] have been investigated. Also, certain systems have been proposed for baggage scanning [41, 45]. E.g., in [41] presents an NQR detector in which the reception is performed with a cryogenically-cooled sensor based on the giant magnetoresistive effect; the system is used to scan luggage for RDX, PETN and ammonium nitrate.

At the commercial level, one can mention the QScan QR 500 detector for luggage scanning [46], manufactured by InVision Technologies (USA). The T-3–03 detector was developed by QRSciences (Australia) [47] for screening envelopes, packages and hand luggage. It can identify the following substances: RDX, PETN and HMX. The NQR-15 detector was developed by Logys (Russia) [48] for screening envelopes or packages; the detection time can take up to 20 s and the equipment can identify several explosives, such as PETN, TNT, RDX and HMX. Anhui Qilootech Photoelectric Technology (China) has produced a detector for scanning envelopes which can detect hydrochloric methamphetamine [49]. The American company SEEQR Security has developed a detector for scanning people and luggage, which allows the identification of explosives urea nitrate, PETN, RDX, TNT, HMX, and ammonium nitrate, in 3 s and using continuous wave excitation [50].

However, existing solutions are developed at the laboratory level and may not meet real operational requirements. At present, there are no commercially available systems on the global market, as there is no up-to-date information regarding them and some manufacturers no longer exist.

3.4 Conclusions

This chapter represented a bibliographic study of methods and equipment for the nuclear quadrupole resonance signal acquisition, in the context of prohibited substances detection.

The advantages of nuclear quadrupole resonance spectroscopy have caused an increased interest for this technique in the field of security. NQR is characterized by very high specificity, entire volume scanning using non-ionizing radiation, applicability to persons, while allowing non-invasive detection, and the fact that the accuracy of detection does not depend on the mass or shape of the substance, but is determined by its chemical structure. These advantages have allowed the use of the technique in the detection of dangerous or contraband substances (luggage scanning, landmine detection), but also in other applications, such as chemical analysis of drugs and minerals, or temperature measurement.

The physical principle of nuclear quadrupole resonance spectroscopy was described, presenting the direct and indirect acquisition methods. Among these, the direct pulse methods are the most utilized, because they increase the signal-to-noise ratio and reduce the scanning time. The main excitation sequences used were detailed, as well as their variants aimed at reducing interference or compensating for the temperature drift of the resonant frequency.

Finally, numerous equipment used in nuclear quadrupole resonance spectroscopy were presented, both at laboratory and commercial level. Research is ongoing in this field, as existing solutions developed at the laboratory level may not meet real operational requirements, and there are currently no commercially available systems on the global market because no up-to-date information is available or certain producers no longer exist.

Appendix

NQR Data for Different Substances

Substance	ν (kHz)	T_2^* (ms)	T_1 (ms)	T_2 (ms)	$d\nu/dt$ (kHz/K)
RDX	5240	0.74	12.3	7.1	−0.47
	5192	1.59	12.6	8.2	−0.43
	5047	0.71	13.3	6.8	−0.43
	3458	0.59	12.1	5.7	−0.33
	3410	0.80	11.1	6.2	−0.06
	3359	0.74	14.6	6.3	−0.27
TNT	870	0.64	4.0	34	−0.109
	859	0.24	3.0	24	−0.223
	848	0.80	9.6	29	−0.151
	844	0.40	4.7	28	−0.121
	842	0.40	3.5	29	−0.181
	837	0.35	2.1	24	−0.122
	768	0.45	9.8	33	−0.190
	751	0.45	2.2	31	−0.241
	743	0.80	3.0	40	−0.148
	740	0.32	5.5	45	−0.169
	714	0.45	4.3	42	−0.094
Ammonium nitrate	497	6.37	14.0	3.5	−0.46
	423	5.31	16.6	14.6	+0.12
Potassium nitrate	567	2.89	20.1	15.5	−0.23
	559	2.89	24.5	14.6	−0.19
	665	0.40	1.9	1.35	−0.19
Heroin	4007	0.26	524	6	–
	3951	0.43	400 ± 100	6 ± 4	–

References

1. Dehmelt, H., Krüger, H.: Pure quadrupole resonance in solids. Naturwissenschaften **37**, 111 (1950)
2. Apostol, M.: Magnetic and Electric Resonance. Cambridge Scholars Publishing (2018)
3. Daniels, D.J.: EM Detection of Concealed Targets Wiley (2010)
4. Zhu, K., Su, T., He, X., Niu, J.: SVD for enhanced explosives detection using NQR. Procedia Eng. **7**, 57–62 (2010)
5. Garroway, A.N. et al.: Remote sensing by nuclear quadrupole resonance. IEEE Tran. Geosci. Remote Sensing **39**(6), 1108–1118 (2001)
6. Deas, R.M., Cervantes, C., Schaedel, S.F.: Landmine Detection by Nuclear Quadrupole Resonance (NQR) (2004)
7. Cardona, L.R.: Nuclear Quadrupole Resonance System for Landmine Detection in Antioquia. Ph.D. thesis, National University of Colombia (2017)

8. Cardona, L., Jiménez, J., Vanegas, N.: Nuclear quadrupole resonance for explosive detection Ingeniare. Revista Chilena de Ingeniería **23**(3), 458–472 (2015)
9. Shao, W., Barras, J., Althoefer, K., Kosmas, P.: Detecting NQR signals severely polluted by interference. Signal Proces. **138**, 256–264 (2017)
10. Yinon, J.: Counterterrorist Detection Techniques of Explosives. Elsevier (2007)
11. Gregorovič, A., Apih, T.: Aplicability of TNT 'super-detection' to multipulse sequences. J. Magn. Reson. **201**(2), 131–136 (2009)
12. Rudakov, T.N.: Detection of Explosives by Nqr Method: Main Aspects for Transport Security, Explosives Detection Using Magnetic and Nuclear Resonance Techniques, pp. 111–138 (2009)
13. Chihara, N. (ed).: Nuclear Quadrupole Resonance Spectroscopy Data, Landolt-Börnstein—Group III Condensed Matter (1997)
14. Kaspar, A.: CONPHIRMER—Project Final Report, Grant Agreement Number 261670. King's College London (2014)
15. Rudakov, T.N.: Magnetic Resonance: NQR Technique and Instruments, Encyclopedia of Spectroscopy and Spectrometry, pp. 700–710 (2017)
16. Monea, C.: The use of nuclear quadrupole resonance spectroscopy for detection of prohibited substances: techniques and equipment. In: 11th International Conference. on Electronics Computers and Artificial Intelligence, 1–6 (2019)
17. Zank, P.A., Apostolos, J.T.: Method and apparatus for sensing the presence of explosives, contraband and other molecules using nuclear quadrupole resonance. Patent US20110187363A1 (2010)
18. Apostolos, J.T., Feng, J., Mouyos, W., McMahon, B.: NQR detection from continuous Rabi transitions. Patent, WO2013049270A1 (2011)
19. Apostolos, J.T., Feng, J., Mouyos, W., McMahon, B.: Using phase matched filters for NQR detection of continuous Rabi transitions. Patent, US8660803B2 (2011)
20. Apostolos, J., Mouyos, W., Feng, J., Chase, W.: Low-power stimulated emission nuclear quadrupole resonance detection system utilizing Rabi transitions. SPIE Proc. **8709** (2013)
21. Apostol, M.: Spectral line of stimulated emission in magnetic resonance. Int. J. Adv. Res. Phys. Sci. **4**(8), 36–41 (2017)
22. Canet, D., Ferrari, M.: Fundamentals of Pulsed Nitrogen-14 Quadrupole Resonance, Explosives Detection Using Magnetic and Nuclear Resonance Techniques, pp. 1–29 (2009)
23. Gregorovič, A., Apih, T.: WURST–QCPMG sequence and 'spin-lock' in 14N nuclear quadrupole resonance. J. Magn. Reson. **233**, 96–102 (2013)
24. Rudakov, T.N.: Modification of SSFP technique for the effective detection of NQR signals. Phys. Lett. A **358**(4), 322–326 (2006)
25. Rudakov, T.N., Belyakov, A.V.: Modifications of the steady-state free-precession sequence for the detection of pure nuclear quadrupole resonance. J. Phys. D: Appl. Phys. **31**(10), 1251–1256 (1998)
26. Garroway, A.N., Miller, J.B., Zax, D.B., Liao, M.-Y.: Method and apparatus for detecting target species having quadropolar muclei by stochastic nuclear quadrupole resonance. Patent, US5608321A (1997)
27. SpinCore Technologies.: RadioProcessor. Internet: http://spincore.com/products/RadioProcessor/. 12 Aug. 2020
28. Tecmag.: Redstone. Internet: http://www.tecmag.com/redstone/. 12 Aug. 2020
29. Tomco Technologies.: BT00250-AlphaS. Internet: http://www.tomcorf.com/assets/standard_pulsed/DS006667A.pdf. 12 Aug. 2020
30. Cardona, L., Miyato, Y., Itozaki, H., Jiménez, J., Vanegas, N., Sato-Akaba, H.: Remote detection of ammonium nitrate by nuclear quadrupole resonance using a portable system. Appl. Magn. Reson. **46**(3), 295–307 (2015)
31. Barras, J., Althoefer, K., Rowe, M.D., Poplett, I.J., Smith, J.A.S.: The emerging field of medicines authentication by nuclear quadrupole resonance spectroscopy. Appl. Magn. Reson. **43**(4), 511–529 (2012)
32. Zhang, X., Schemm, N., Balkir, S., Hoffman, M.W.: A low-power compact NQR based explosive detection system. IEEE Sensors J. **14**(2), 497–507 (2014)

33. Rudakov, T.N., Rayner, T.J., Hayes, P.A., Russeth, K.L.: Detection of explosives by quadrupole resonance method: New aspects for security, Detection and Disposal of Improvised Explosives, pp. 191–204
34. Lužnik, J., Pirnat, J., Jazbinšek, V., Apih, T., Blinc, R., Seliger, J., Trontelj, Z.: Improved 14N nuclear quadrupole resonance detection of trinitrotoluene using polarization transfer from protons to 14N nuclei. J. Appl. Phys. **102**(8), 084903 (2007)
35. Glickstein, J., Mandal, S.: An automated instrument for polarization-enhanced broadband nuclear quadrupole resonance (NQR) spectroscopy. Rev. Sci. Instrum. **89**(9), 093106 (2018)
36. Begus, S., Pirnat, J., Jazbinsek, V., Trontelj, Z.: Optical detection of low frequency NQR signals: a step forward from conventional NQR. J. Phys. D Appl. Phys. **50**(9), 095601 (2017)
37. Monti, M.C., Alexson, D.A., Okamitsu, J.K.: NQR detection of explosive simulants using RF atomic magnetometers. In: Detection and Sensing of Mines, Explosive Objects, and Obscured Targets XXI (2016)
38. Lee, S.K., Sauer, K.L., Seltzer, S.J., Alem, O., Romalis, M.V.: Subfemtotesla radio-frequency atomic magnetometer for detection of nuclear quadrupole resonance. Appl. Phys. Lett. **89**(21), 214106 (2006)
39. TonThat, D.M., Clarke, J.: Direct current superconducting quantum interference device spectrometer for pulsed nuclear magnetic resonance and nuclear quadrupole resonance at frequencies up to 5 MHz. Rev. of Sci. Instrum. **67**(8), 2890–2893 (1996)
40. He, D.F., Tachiki, M., Itozaki, H.: N-14 NQR using a high-Tcrf SQUID with a normal metal transformer. Supercond. Sci. Technol. **21**(1), 015023 (2007)
41. Poullain, D., et al.: Research for the Detection of Explosives at CEA: Towards Operational Use, Future Security, pp. 412–427 (2012)
42. Peshkovsky, A.S., Cattena, C.J., Cerioni, L.M., Osán, T.M., Forguez, J.G., Peresson, W.J., Pusiol, D.J.: Noise-resilient multi-frequency surface sensor for nuclear quadrupole resonance. J. Magn. Reson. **194**(2), 222–229 (2008)
43. Scharfetter, H.: An electronically tuned wideband probehead for NQR spectroscopy in the VHF range. J. Magn. Reson. **271**, 90–98 (2016)
44. Scharfetter, H., Bödenler, M., Narnhofer, D.: A cryostatic, fast scanning, wideband NQR spectrometer for the VHF range. J. Magn. Reson. **286**, 148–157 (2018)
45. Mozzhukhin, G.V., et al.: The Detection of Industrial Explosives by the Quadrupole Resonance Method: Some Aspects of the Detection of Ammonium Nitrate and Trinitrotoluene, Explosives Detection Using Magnetic and Nuclear Resonance Techniques, pp. 231–244 (2009)
46. InVision Technologies.: QScan QR 500. Interenet: http://www.tepsa.com.ar/pdf/invision/QR500.pdf. 12 Aug. 2020
47. Rudakov, T.N.: Further Improvement of NQR Technique for Detection of Illicit Substances, Magnetic Resonance Detection of Explosives and Illicit Materials, pp. 3–17 (2013)
48. Logys.: NQR-15. Internet: http://www.varus.mx/english/catalogue/71/248/. 24 Sep. 2018
49. Anhui Qilootech Photoelectric Technology.:, NQR small-scale postal channel. Internet: www.qilootech.net/nqr_small_scale_postal_channel_p6839.html. 24 Sep. 2018
50. SEEQR Security.: SEEQR. Internet: https://jon-bell-742j.squarespace.com/seeqr. 24 Sep. 2018

Chapter 4
Signal Processing and Analysis Techniques Applied in Nuclear Quadrupole Resonance

4.1 Classification of the Signal Processing Techniques

The signal processing and analysis (SPA) techniques applied in NQR detection generally have the following purposes: (1) increase the SNR and (2) extract (recognize) the response signal. Also, they usually aim to extract useful data from the NQR signal and to reduce interference. A classification has been proposed based on multiple criteria [1]:

- Type of pulse excitation method used: classical NQR (cNQR) and stochastic NQR (sNQR);
- Number of sensors used: single sensor (single channel), multi-sensor (multi-channel);
- Location of implementation within the reception (RX) chain: pre-, post-processing;
- Excitation method applied: continuous wave techniques, pulse techniques.

In the following text, the terms detector, technique, solution and algorithm are used interchangeably. Figure 4.1 classifies the techniques based on the location of implementation.

The main difference between the pre- and post-processing solutions is that the first improve the signal-to-noise ratio during the acquisition of the NQR signal, whereas the second are applied to the pre-processed signal to extract the response signal and its features [2]. Signal averaging and quadrature detection are the most utilized pre-processing solutions.

The majority of post-processing and analysis (PostPA) solutions were designed for classical NQR since this what the first detection type discovered. sNQR detection has been proposed afterwards and gradually became the focus of researchers due to its advantages, such as low-power detection, the possibility to scan people and the simplified construction of portable detectors used, e.g., in landmine detection [3].

C. Monea and N. Bizon, *Signal Processing and Analysis Techniques for Nuclear Quadrupole Resonance Spectroscopy*, Signals and Communication Technology,
https://doi.org/10.1007/978-3-030-87861-0_4

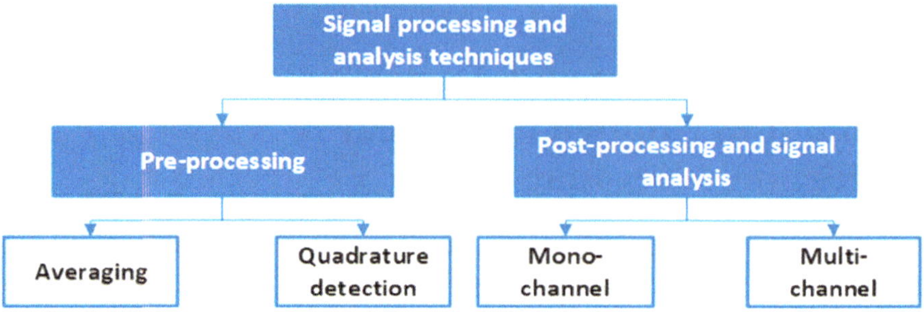

Fig. 4.1 SPA techniques classification based on the location of implementation

A large percentage of the PostPA solutions is intended for single-channel applications based on pulse excitation. This is due to the lower cost of this type of detectors, as well as the advantages of pulse excitation.

The following is a classification of PostPA techniques. The number of channels was selected as the classification criterion, because the majority of algorithms fall only into one of the two possible categories, while the other criteria present several overlaps. Also, this classification highlights the differences regarding the algorithms' complexities for single- and multi-channel detection. Figure 4.2 presents the single-channel processing techniques.

The techniques fall into several main categories. For clarity, they are illustrated in the order they were developed. There are six categories, ranging from spectral

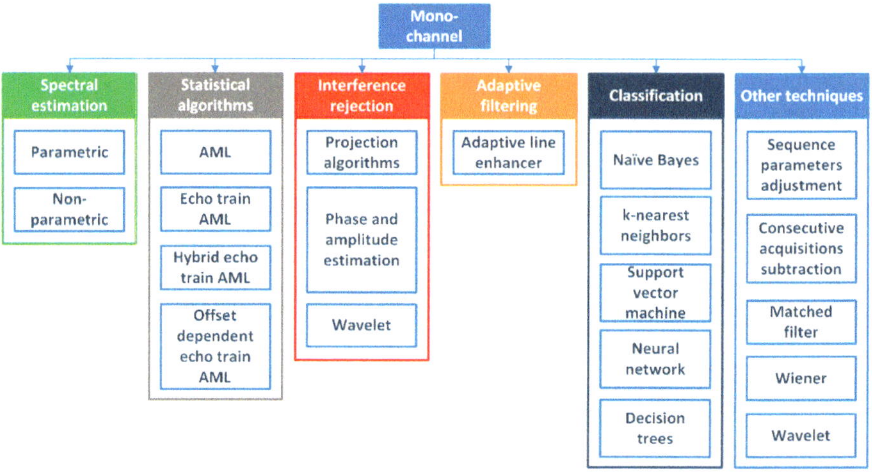

Fig. 4.2 PostPA techniques applied in single-channel NQR detection [1] (Reprinted from J. of the Franklin Inst., vol. 357, issue 17, C. Monea, A review of NQR signal processing and analysis techniques, Pages 13,085–13,124, Copyright 2020, with permission from Elsevier [OR APPLICABLE SOCIETY COPYRIGHT OWNER])

estimation solutions to classification algorithms. The former were developed as an improvement for the classical energy detectors that had poor performance in case of low SNR. Their improvement of the was driven by the need to extract and use more information from the NQR signal. The interference rejection algorithms were proposed to improve the SNR and usually perform the detection. The adaptive filtering category is indicated to highlight the techniques adapted for NQR detection. The classification solutions aim to recognize the response signal based on its features. The algorithms presented in the last category are proposals that improve detection in particular conditions.

From a chronological point of view, the first technique used was the discrete Fourier transform (DFT) that performed evaluated the amplitude of the response's spectral peak. This spectral estimation technique is also found in the literature as DMA (demodulated approach), correlated with quadrature detection, or as FT-NQR (Fourier transform NQR). It allows the detection of a single resonant frequency, which driven the research of other techniques that can extract and use several characteristics of the NQR signal. Other categories had a relatively separate development from the first ones, and certain results obtained in single-channel detection were extended to multi-channel applications (e.g., EPIC-DEPIC or REMIQS-ESPIRE).

ETAML uses previously available information regarding the response signal and represents an improvement of the existing AML solution. It performs a generalized likelihood ratio test and compares it with a predetermined threshold value with the aim of finding the false alarms probability that is acceptable for the detection application. Frequency-selective versions (FSAML and FETAML) have been developed for these solutions, which process the data located in the frequency range where the NQR signal is located. All these algorithms are based on complete knowledge of the complex signals' amplitudes up to a multiplicative constant, but in practice, they are influenced during the measurement (scanning) by several factors. Thus, the RETAML algorithm was developed, which integrates a priori information about the complex amplitudes and allows uncertainties regarding this information. Subsequently, it was developed to obtain the frequency selective version FRETAML. LSETAML differs from ETAML in the sense that it considers the complex amplitude vector completely unknown (no a priori information) and estimates it. The FLSETAML version was also developed.

Starting from a signal model for polymorphic substances (which have several forms of crystallization), the ETAML algorithm was further improved. Thus, the hybrid versions F-/HETAML were proposed. The latter has been improved to allow uncertainties regarding the assumed signal amplitudes, resulting REMIQS.

The SEAQUER algorithm was developed to attenuate interference in stochastic NQR experiments by operating on a noise subspace and using the signal pre-whitening technique; however, it requires estimates of the attenuation constants (a priori information). The ETCAPA algorithm is based on a model of the entire pulse train and has been proposed for estimating the attenuation constants of each signal component from an unknown substance, while allowing RFI attenuation. ET-ESPRIT is an alternative method for estimating the parameters that is also based on this model. It does not require a priori information regarding the parameters' values, but only the

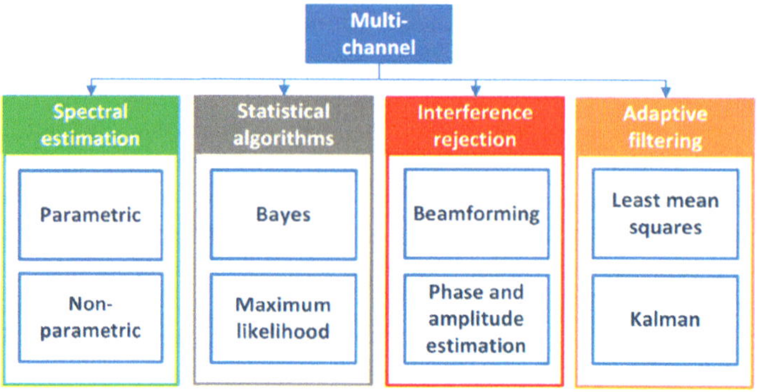

Fig. 4.3 PostPA techniques applied in multi-channel NQR detection [1] (Reprinted from J. of the Franklin Inst., vol. 357, issue 17, C. Monea, A review of NQR signal processing and analysis techniques, Pages 13,085–13,124, Copyright 2020, with permission from Elsevier [OR APPLICABLE SOCIETY COPYRIGHT OWNER])

number of spectral lines. Subsequently, the SEAQUER algorithm was generalized, obtaining RESPEQ and REWEAL. The interference attenuation algorithms based on Fourier and wavelet analysis have been proposed, and WICAML can be mentioned.

The classification solutions have been recently developed for detection, and algorithms such as the neural network or the support vector machine have been investigated.

Figure 4.3 illustrates the signal processing solutions designed for multi-channel detection.

These are presented similarly to the techniques for single-channel applications. These smaller number of solutions can be explained by the fact that multi-channel detectors are more complex and costly than single-channel ones. The existence of multiple antennas allows the application of spatial filtering (diversity) algorithms.

Some algorithms are intended for both types of detection (cNQR and sNQR), such as EPIC, ESPIRE or NORRDIQ. Other techniques were initially developed for one type of detection and adapted for the other case, such as SEAQUER and RCDAML, which were initially designed for sNQR and later adapted for cNQR, obtaining c-SEAQUER and RTDAML, respectively.

The classification of the SPA algorithms applied in NQR can be found in Appendix 1. In order to provide a more complete picture, Appendices 2 and 3 present the development timeline of the algorithms, respectively the links between them. Also, in Appendix 4 one can see the distribution of post-processing techniques according to the type of detection.

4.2 Pre-Processing Techniques

The main disadvantage of nuclear quadrupole resonance is the very low signal-to-noise ratio. Therefore, the challenges of the processing techniques are to reduce interference and extract the response signal from noise [4]. In both continuous wave and pulse NQR, the signal is digitized using an analog-to-digital converter, and then it is processed. The pre-processing solutions are used to improve the signal-to-noise ration during signal acquisition. In other words, they process the signal during its acquisition, and the most used techniques are:

- Signal averaging;
- Quadrature detection.

There are also processing techniques implemented in the analog domain (signal conditioning techniques), such as adaptive filtering based on the continuous time LMS algorithm (CTLMS) to reduce RFI [5], but it is not the focus of this work.

4.2.1 Signal Averaging

Signal averaging is applied to extract (separate) a repetitive signal from noise while avoiding distorting it [6]. From a spectral point of view, the noise may overlap or differentiate from the desired signal. In the first case, the use of a filter to eliminate noise would distort the signal, and the signal averaging technique would be useful to increase the SNR. This is effective if the noise is random. Each acquisition is made at the same time as the previous ones so that the samples of the new scan are added to the previous ones.

Considering that the averaged signal contains noise, $n(t)$, and the useful (desired) signal, $s(t)$, then after sampling and storing the samples in memory, the following signal is obtained:

$$\sum_{k=1}^{N} f(iT) = \sum_{k=1}^{N} s(iT) + \sum_{k=1}^{N} n(iT), i = 1, \ldots, m \tag{4.1}$$

where i is the memory location where the sample i is stored, T is the sampling period, m is the number of samples, and N is the number of repetitions (scans). The useful signal corresponding to sample i is identical at each average (iteration) if it is stable and the acquisitions are synchronized over time. Thus, it can be written:

$$\sum_{k=1}^{N} s(iT) = Ns(iT) \tag{4.2}$$

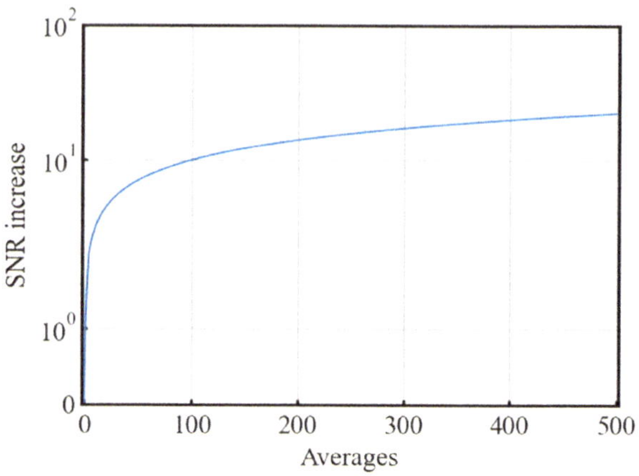

Fig. 4.4 The increase of the signal-to-noise ratio with the number of averages

By considering that the useful signal and the noise are uncorrelated, and the noise is random, with zero average value, then after N iterations, the RMS value (standard deviation) of the noise will be σ_n:

$$\sum_{k=1}^{N} n(iT) = \sqrt{N\sigma_n^2} = \sqrt{N}\sigma_n \tag{4.3}$$

The signal-to-noise ratio is obtained by dividing Eqs. (4.2) and (4.3):

$$SNR_N = \frac{Ns(iT)}{\sqrt{N}\sigma_n} = \sqrt{N}SNR \tag{4.4}$$

Equation (4.4) shows that signal averaging increases the SNR by $\sqrt{N}$, as illustrated in Fig. 4.4.

Signal averaging can be considered a numerical filtering operation using a comb filter that adds delayed versions of the input signal with itself, causing coherent components to increase and incoherent ones to decrease.

During signal averaging, the desired signal will add coherently, and the noise incoherently. As shown in Fig. 4.5, after a certain number of acquisitions, the desired signal will be identifiable from the received one.

4.2.2 Quadrature Detection

Initially, quadrature detection was applied in NMR by utilizing two coils placed at an angle of 90°, and was later used in NQR. The technique is utilized to distinguish the response signal's components with frequencies lower or higher than the excitation frequency. The block diagram of a quadrature detector is illustrated in Fig. 4.6.

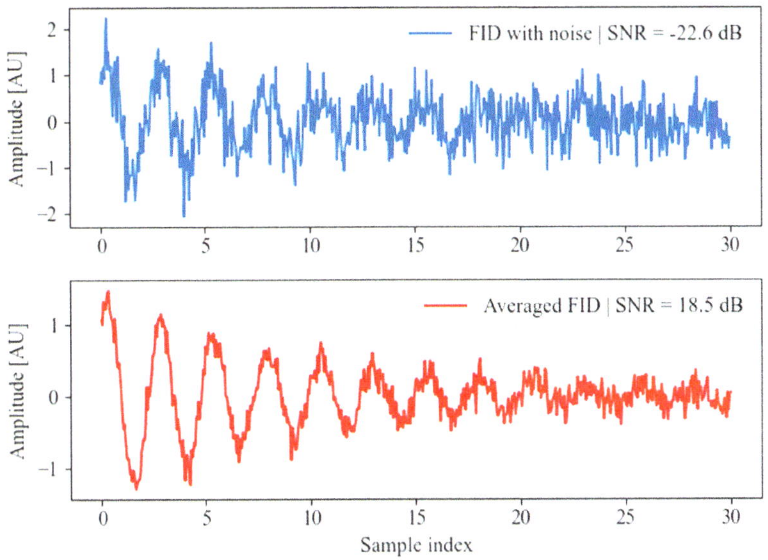

Fig. 4.5 FID signal improvement by performing 16 averages (AU—arbitrary units)

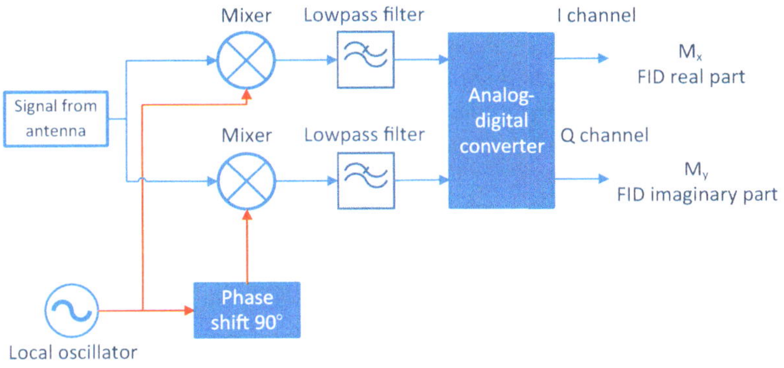

Fig. 4.6 Block diagram of a quadrature detector

The response signal from the antenna (coil) is provided to the two mixers, corresponding to the real (I; cosine) and imaginary (Q; sine) channels. The reference signal is ensured by a local oscillator, with the mention that the signal supplied to the Q channel is phase-shifted by 90°. After filtering, the x and y components are digitized and transmitted to the modules responsible for signal averaging and SPA. The components with lower or higher frequencies than the excitation frequency can be graphically visualized using the complex discrete Fourier transform. If two detection coils are used, the SNR will increase by $\sqrt{2}$, but when using a single coil, the SNR remains unchanged [7].

Also, quadrature detection spectrally limits the noise by low pass filtering the signal. This can be performed either in the digital domain, by positioning the analog-to-digital converter in front of the detector module, or in the analog domain, as shown in Fig. 4.6. Also, the filtering can be done with or without decimation to reduce the amount of acquired data. Although it offers advantages, there are situations in which it is desired to acquire the raw signal instead of the signal resulting from the quadrature detection, in order to have a larger amount of data for SPA.

4.3 Post-Processing and Analysis Techniques

The PostPA solutions process the pre-processed signal and extract it for evaluation and interpretation purposes. The initial solutions were based on the DFT, but their poor performance in low signal-to-noise conditions has driven the research of techniques that extract the NQR response signal's features and use theoretical signal models. The first solutions were proposed for cNQR; later, they were developed for sNQR detection. Generally, the recent algorithms have the following characteristics [8]:

- They exploit the signal's structure;
- They exploit the spectral lines' temperature dependence;
- They provide robustness to uncertainties regarding the estimated amplitudes;
- They exploit the materials' polymorphic structure;
- They detect multiple polymorphic forms;
- They identify substance mixtures;
- They attenuate radio frequency interference.

For clarity, the post-processing and analysis techniques are be presented separately depending on the type of detection, in single-channel and multi-channel techniques, respectively, and each section describes the techniques from the sub-categories presented above.

4.3.1 Single Channel Detection Techniques

Power spectrum estimation has been used in NQR and can be parametric and non-parametric; the former typically achieves a better resolution spectrum [9]. Statistical processing is applied by using the Fourier transform of the autocorrelation function of random processes, considered stationary in a broad sense. The parametric techniques are based on a signal model developed using previously available information. They involve estimating the model's parameters from the current data and replacing them in the model-specific power spectral density formula.

In [10] several algorithms are proposed that aim to evaluate the average signal strength based on parametric and non-parametric estimators, in order to detect landmines. The free induction decay signal is represented as a sum of (damped) sinusoids:

$$y(t) = \sum_{k=1}^{d} \alpha_k e^{(i\omega_k - \beta_k)t} + w(t) \tag{4.5}$$

where:

- α_k is the amplitude for the kth component;
- ω_k is the frequency for the kth component;
- β_k is the attenuation constant for the kth component;
- w is additive colored noise;
- d is the number of resonance lines.

Also, the echo signal is modeled as two symmetric FID signals:

$$y(t) = \sum_{k=1}^{d} \alpha_k e^{i\omega_k t - \beta_k |t - t_0|} + w(t) \tag{4.6}$$

where t_0 is the center of the echo.

The following non-parametric solutions were used: DFT and periodogram. Also, the following parametric techniques were applied: autoregressive (AR) models (covariance, Yule-Walker, modified covariance and Burg), and the MUSIC method (multiple signal characterization). The evaluation of spectral estimation techniques was performed by simulating and analyzing 180 g of TNT, and showed that the AR technique based on the modified covariance offers the best performance.

The use of non-parametric estimators, such as Capon or APES, for estimating the echo/-train damping constants does not allow detailed estimation of the echo train's (ET) structure. The ETCAPA algorithm has been proposed as a solution [3]. The application of the Capon, CAPES and CAPA spectral estimation techniques was described in [11]. These were tested by Monte Carlo simulation, in the case of methamphetamine and TNT. The results of the study indicated that all algorithms accurately estimate the resonance frequency, each of which is appropriate for specific scenarios.

In [12] the spectral estimation based on the ESPRIT method applied to the ET is proposed. The ET-ESPRIT's model does not need to know its parameters' values, but only of the number of spectral lines.

The AML algorithm compensates the spectral lines' temperature dependence [13]. It compares the measured spectral lines, whose temperature dependence is known, with data obtained from the estimated model. AML evaluates the spectral peak corresponding to the best-fitting temperature. The model is given by the equation:

$$y(t) = \sum_{k=1}^{d} \alpha_k e^{-\beta_k |t - t_0| + i\omega_k(T)t} + w(t) \tag{4.7}$$

where $t = 0, 1, \ldots, \mathrm{N} - 1$, k is the spectral line, t_0 is the offset of the known spectral peak. $\omega_k(T)$ represents the frequency shift function with the temperature for the kth spectral line, defined as follows:

$$\omega_k(T) = a_k - b_k T \tag{4.8}$$

where $k = 1, \ldots, d$, a_k and b_k are constants with known values.

The relative ratio of the amplitudes is known for a given substance. In order to exploit this knowledge, the following equality is considered:

$$\alpha_k = \rho K_k \tag{4.9}$$

where ρ and K_k are scaling constants.

The algorithm is compared to the classical DFT-based detectors (DMA) using simulated and real data and is significantly more efficient than DMA, especially if the temperature error is considered.

The FSAML algorithm is a frequency-selective version of AML, which operates with a subset of available frequencies, which gives it the advantages of increased robustness to narrowband RFI and decreased computational complexity [14]. The increased robustness to interference is obtained at the cost a slightly lowered detection performance. In addition, the calculation complexity is about 3.5 times lower than AML.

Somasundaram et al. have suggested in [15] to process the individual responses as a sequence instead of averaging them. Thus, the AML and FSAML solutions were extended and the ETAML and FETAML algorithms were obtained. They process the pulse train that is not averaged, in order to exploit the information regarding the echo decay over time, leading to an increase in the probability of detection. The algorithms were evaluated using both simulated and real data. The data sets contain 32 echoes consisting of 256 samples, measured at 180 g TNT, in laboratory conditions, without noise and with noise having the amplitude 18 dB higher than that of the useful signal, respectively. F-/ETAML offers the highest detection probabilities in both situations.

The algorithms presented so far need to know the complex amplitudes up to a multiplicative constant (see Eq. 4.9). But in real situations there may be variations in the response signal's amplitude. For example, in the case of landmine detection, the distance between the substance and the detection coil may differ between targets; therefore, using the same excitation sequence and power, the rotation angle of the quadrupolar nuclei will vary, causing variations in the amplitude of the received signal. Some solutions to this problem are proposed in [16].

The model used is described by the following equation which represents the mth echo in an echo train:

$$y_m(t) = \sum_{k=1}^{d} \alpha_k e^{(-\eta_k(t+2mt_{sp})-\beta_k|t-t_{sp}|+i\omega_k(T)t)} + w_m(t) \tag{4.10}$$

where $m = 0, 1, \ldots, M-1$, $t = t_0, \ldots, t_{N-1}$, η_k is the attenuation constant of the echo train corresponding to the kth spectral line, the number of echoes is given by M, t_{sp} represents the time between the initial and refocusing pulse, with the echo's peak located at t_P.

LSETAML is a variant of the ETAML algorithm that estimates the complex amplitudes without requiring information about them. RETAML normalizes the amplitude vector and searches in an uncertainty region (hypersphere) around a specified vector. It can be assumed that the temperature of the substance is within a generally known range of $\pm$ 5 K around the reference value. Based on this hypothesis, the frequency-selective versions FRETAML (frequency selective ETAML) and FLSETAML (frequency selective LSETAML) were developed. According to the evaluation, the robust and frequency-selective algorithm provides the best performance.

The F-/HETAML algorithms were developed to exploit signals from a mixture in order to improve the detection probability [17]. Also, the hybrid solutions can estimate the relative proportions of the components in the mixture. They can also be applied to polymorphs. The data model used is described in Eq. (4.10), except that it is specific to each individual polymorph that is part of the substance. The algorithms were evaluated under laboratory conditions on 500 g of TNT, consisting of two polymorphic forms, monoclinic and orthorhombic, and achieved better performance than the non-hybrid versions.

Butt et al. propose in [18] to generalize the hybrid solutions in order to allow amplitudes uncertainties, which can occur in practice. Thus, the LS-FHETALM and REMIQS solutions were proposed. LS-FHETALM is a generalized variant of FHETAML that allows uncertainties regarding the complex amplitudes. REMIQS, first estimates the polymorphs' relative proportions. Afterwards it calculates the relative amplitudes associated with the mixture's substances. The algorithms were evaluated under the conditions presented above and both performed better than the non-robust form. Also, REMIQS achieves the best detection probability.

When the excitation frequency does not perfectly match the substance's temperature, then the SLSE response can have an amplitude modulation due to off-resonance effects [8]. The F-/ODETAML algorithms can exploit the dependence of the response signal's amplitude to the off-resonance effects, by using a particular data model.

Regarding **interference rejection**, in [19, 20] two algorithms are proposed for reducing interference in sNQR experiments. The first assumes the RFI to be found in its own subspace, that can be estimated from signal of interest-free (SOI) samples. Then the subspace is used to generate a matched detector. The algorithm is called SEAQUER (subspace-based evaluation of quadrupole resonance signals exploiting robust methods).

sNQR relies on correlating the RX signal with a pseudo-/random excitation sequence [19]. RCDAML exploits the result from the correlation with the signal without the samples of interest, in order to determine the noise covariance matrix.

The solutions were tested using a dataset composed of 1000 acquisitions performed in laboratory conditions on 180 g monoclinic TNT and with the addition of noise with a bandwidth of 25 kHz and an interference-to-signal ratio (ISR) of more than 30 dB. SEAQUER offers the best detection probability, as well as the highest interference rejection.

SEAQUER and RCDAML have also been adapted to cNQR detection [21]. In this case, the interference subspace is estimated based on the samples without the SOI, obtained when the refocusing pulses are absent (called "echoes" without SOI). The proposed algorithms are named c-SEAQUER and RTDAML, respectively. In the case of an ISR of 60 dB, it is necessary to acquire at least 500 echoes without the signal of interest in order for the proposed algorithms to become efficient.

Similarly, c-/SEAQUER were generalized for polymorphic forms, resulting the RESPEQ algorithm [22]. Compared to SEAQUER and REMIQS, the proposed algorithm offers the highest probability of detection at an average ISR of 40 dB.

The REWEAL algorithm is proposed in [23], for the case when the signal of interest is affected by narrowband RFI and broadband noise is considered. This solution searches for the RFI and SOI in hypercones around the signal subspaces. REWEWAL can be applied if the amount of secondary data is small.

Swärd investigates in [24] the reduction of stationary interference by estimating them from secondary data, even when their frequencies overlap with those of the signal of interest. The proposed algorithm, called EPIC, firstly estimates the dominant spectral components from the secondary data, and then subtracts them from the primary data sets (where the SOI is contained). EPIC was used in conjunction with the RETAML algorithm to detect the NQR signal. However, EPIC requires that secondary data to be synchronous with the principal data, which is hard to obtain in practice [25].

Shao et. al. proposes in [26] a solution to cancel the interference that overlaps the useful signal, in case of landmine detection. The solution was used in conjunction with AML for detection, and is called interference cancellation approximate maximum likelihood (ICAML). The RFI components are estimated using a cost function and and Fourier analysis for stationary ones and wavelet analysis for non-stationary ones, respectively. Unlike previous solutions, this algorithm does not require secondary data for RFI estimation.

The F-/ETAML algorithm is developed by adding a solution to attenuate interference close to the spectral bands of NQR responses, thus obtaining the interference cancellation (selective frequency) echo train approximate maximum likelihood (ICETAML/ICFETAML) [25]. They offer better performance than previous solutions.

Shao et al. propose in [27] a solution for rejecting non-stationary interference overlapping the NQR signal bands, without affecting the response signal. The proposed algorithm, called WICAML, uses the Gabor-Morlet wavelet analysis for estimating the time-varying RFI and performs better than Fourier analysis-based solutions, either

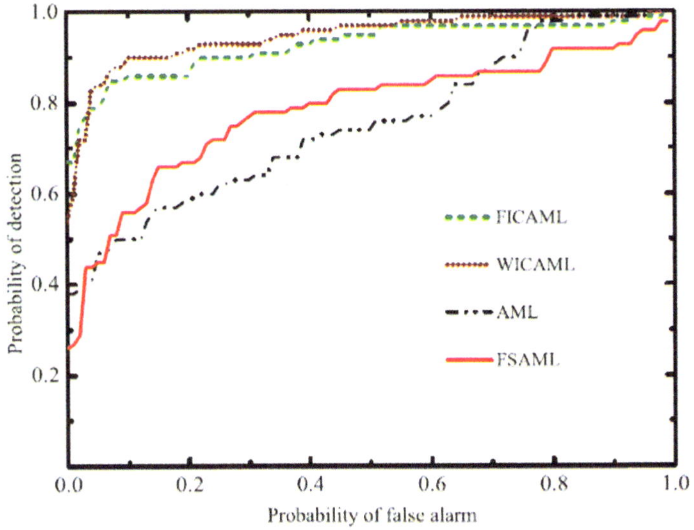

Fig. 4.7 ROC curves for the WICAML, FICAML, AML and FSAML algorithms [27] (Reprinted from Sig. Proc., vol. 154, W. Shao, J. Barras, P. Kosmas, A novel wavelets method for cancelling time-varying interference in NQR signal detection, Pages 238–249, Copyright 2019, with permission from Elsevier [OR APPLICABLE SOCIETY COPYRIGHT OWNER])

frequency-selective or adaptive. Basically, the algorithm first cancels the interference, then it applies the AML solution to perform the detection. The solution was tested under conditions similar to the previous ones and, as seen in Fig. 4.7, it has the best detection probability. FICAML (Fourier analysis based ICAML) is, in fact, ICETAML.

Adaptive filters have been applied in NQR to reduce noise [28, 29]. Specifically, the use of an ALE was proposed decorrelate the noise and keep the useful signal correlated.

The **classification solutions** are proposed in [30] to discriminate the signal generated by ammonium nitrate landmines from RFI and noise, based on several characteristics such as: amplitude of the spectral component at the reference frequency, frequency of the higher spectral peak in a frequency band of interest, etc. The following classification algorithms were investigated: k-nearest neighbors (k-NN), naive Bayes, neural network, support vector machine (SVM) and a collection of classifiers (decision trees; EC - ensemble of classifiers). The latter offered the best detection probability, achieving 85.7% in real conditions.

The use of the neural network was proposed in [31] for performing sNQR detection. The robustness of the algorithm variations of the NQR signal's properties in real scenarios and the good performance in the presence of interference is confirmed, surpassing the traditional signal processing and estimation methods.

Other techniques that cannot be included in the categories described above were applied in single-channel detection. In [32, 33] an adaptive solution was described to adjust the excitation pulse duration based on the peak-to-peak value of the average of N response signals obtained with a fixed pulse width.

Yang et al. propose in [34] to use a Wiener filter composed of multiple stages to reduce noise and interference. In the case of stimulated emission, a matched filter has been proposed for NQR detection [35]. It considers the phase differences introduced by the signal paths related to the reference and the received signals, respectively; it accepts or rejects the received signal depending on these phase differences.

NQR detection was performed by using wavelet analysis, as described in [36]. The solution calculates the mutual wavelet spectrum (mutual coherence) between a reference signal and the received response. The level of coherence is used to discriminate the NQR response.

4.3.2 Multi-channel Detection Techniques

The existence of more than one antenna enables the use of better and more complex RFI rejection solutions. The main challenges to overcome are the synchronization of the main and secondary data, as well as the compensation of differences between the channels (phase, gain).

In [37], the use of the following **spectral estimation** solutions is proposed: MUSIC and periodogram. Also, the least mean squares algorithm is applied to adaptively reduce the RFI. An array of antennas is proposed in [38] to exploit temporal and spatial information with the aim of improving detection. The discrimination based on the Bayes criterion or algorithm was proposed in [39] to identify the NQR response signal for landmine detection. The difficulties of detection are given by the fact that the resonant frequency is unknown due to its dependence on the ambient temperature, as well as the low SNR and high interference level. Jiang et al. investigate in [38], the use of an antenna array to estimate the response amplitude. The MLH and M^3L solutions are applied.

In the case of multi-channel detectors, **interference rejection** is performed using spatial filtering (diversity) techniques. In [40, 41] the SCB, RCB and APES beamformers are applied to cancel correlated RFI. This correlation is further used in [42] for improving the detection of TNT. In [43], the use of the DCRCB-GLRT solution is described for joint detection of TNT and RDX. In [44], the ETAML solution was extended for multi-channel detectors. The new algorithm, called BICETAML, ensures robustness against the gain differences between the channels.

The ALS technique is investigated in [38]. Simulation studies show that this technique is superior to M^3L, but has lower performance when the signal of interest and interference are very close in terms of frequency.

Two techniques are proposed in [45] for detecting in interference conditions: NLS and FSMC. The first works throughout the entire range of frequencies and is sensitive to amplitude modulated radio transmissions.

In [46, 47] the FSMC algorithm is generalized to detect mixtures of substances (polymorphs). The new hybrid solution, called ESPIRE, is a variant of REMIQS and allows a temperature uncertainty of ± 10 K.

Although it has outperformed the algorithms presented above, ESPIRE has a significant limitation. The technique is based on the MLH estimation of the spatial covariance matrix and the NQR signal parameters. These estimates are made on the assumption that the noise is white. But the performance of the algorithm is affected in the presence of correlated noise and interference. Thus, in [48] the aim is to reduce this problem by attenuating the correlated RFIs before estimating the MLH. The NORRDIQ solution is proposed as an extension of ESPIRE. Similarly, DEPIC is developed as an extension of EPIC [49]. DEPIC also tracks the gain differences between the channels.

Regarding **adaptive filtering**, the least mean squares algorithm is used in [37] to reduce radio frequency interference. An interference attenuation of 50 dB has been reported using a 1-tap filter. Tan et al. propose in [50] the use of a Kalman filter to detect the presence of the NQR response in the signal resulting after interference rejection.

4.4 Future Research Directions

Given the development timeline of the techniques applied in NQR detection, it must be mentioned that new research can aim to develop solutions that are more robust to the noise and uncertainties regarding the model's parameters, as well as SPA algorithms, such as those specific to machine learning (ML).

ML is a field of research with multiple applications and numerous articles published every year [51]. There are dozens of algorithms developed in this field, the most efficient being those based on deep architectures (deep learning techniques): deep neural networks (DNN), deep belief networks (DBN), autoencoders (AE), convolutional neural networks (CNN), recurrent neural networks (RNN) and others [52]. Better algorithms appear every year, among which the following can be mentioned: LeNet, AlexNet, VGG, GoogLeNet, ResNet [53]. Machine learning also offers many benefits, such as generating new features and unsupervised learning [54]. In particular, it includes techniques useful for signal analysis in low SNR conditions, such as NQR detection, and has been shown to perform well in signal recognition applications [55].

Machine learning solutions were proposed for NQR applications [30, 31], but the proposed algorithms represent only a small part of the available techniques and can be replaced by more recent solutions, such as those previously mentioned. The fusion of an algorithm with better feature extraction techniques can provide very good results in the analysis of NQR signals. A machine learning solution for NMR analysis is proposed in [56]. Although it is not applied in NQR experiments, a similar approach can be used in this case as well, as suggested in the diagram in Appendix 1.

4.5 Conclusions

This chapter aimed at perform a literature review of the SPA techniques used in NQR applications, with the aim of understanding the principles, performance and application of the techniques (1), to allow the selection of the best performing techniques (2) and to compare these algorithms (3) in order to propose new solutions (4).

The SPA techniques were classified and described in separate categories. Also, the development timeline of the algorithms and the relationships between them were detailed.

Signal averaging is practically indispensable in applications based on pulse excitation and is the most used technique to increase the signal-to-noise ratio and decrease the scanning time. Quadrature detection is widely used because it allows the visualization of the frequency deviation from the excitation frequency.

The PostPA solutions have undergone a great development, starting from those based on the signal's energy and reaching the classification solutions. The statistical techniques estimate the (assumed) signal model's parameters. The performance of RFI rejection techniques has been improved by considering mixtures and interference subspaces. The adaptive solutions involved the use of different filters for noise reduction. Multi-channel solutions use several antennas, which allows them to attenuate interference more efficiently by using adaptive beamforming algorithms.

The classification solutions have been recently proposed. Good results were obtained regarding the identification of ammonium nitrate landmines and in stochastic detection, in conditions of very low signal-to-noise ratio, where it is not feasible to compare the spectral amplitude of the response with a predefined threshold, because false alarms may occur. Several classification techniques were investigated such as k-nearest neighbors, naive Bayes, neural network, support vector machine and a collection of classifiers.

The solutions for signal classification used so far represent only a small part of the available techniques. Thus, their replacement with recent solutions is proposed. In addition, the merge of a machine learning algorithm with better feature extraction techniques is proposed, as these could provide very good results in the analysis of NQR signals.

Appendix

Signal Processing and Analysis Techniques Classification

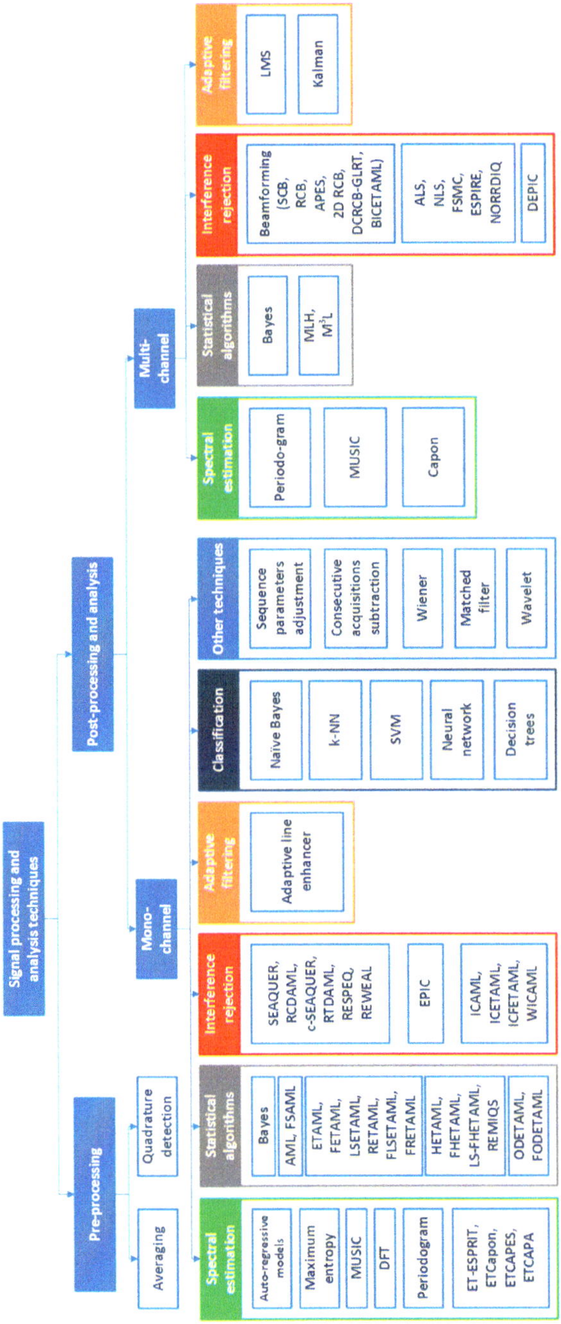
Signal processing and analysis techniques
Pre-processing
Averaging
Quadrature detection
Post-processing and analysis
Mono-channel
Spectral estimation
Auto-regressive models
Maximum entropy
MUSIC
DFT
Periodogram
ET-ESPRIT, ETCapon, ETCAPES, ETCAPA
Statistical algorithms
Bayes
AML, FSAML
ETAML, FETAML, LSETAML, RETAML, FLSETAML, FRETAML
HETAML, FHETAML, LS-FHETAML, REMIQS
ODETAML, FODETAML
Interference rejection
SEAQUER, RCDAML, c-SEAQUER, RTDAML, RESPEQ, REWEAL
EPIC
ICAML, ICETAML, ICFETAML, WICAML
Adaptive filtering
Adaptive line enhancer
Classification
Naïve Bayes
k-NN
SVM
Neural network
Decision trees
Other techniques
Sequence parameters adjustment
Consecutive acquisitions subtraction
Wiener
Matched filter
Wavelet
Multi-channel
Spectral estimation
Periodo-gram
MUSIC
Capon
Statistical algorithms
Bayes
MLH, M³L
Interference rejection
Beamforming (SCB, RCB, APES, 2D RCB, DCRCB-GLRT, BICETAML)
ALS, NLS, FSMC, ESPIRE, NORRDIQ
DEPIC
Adaptive filtering
LMS
Kalman

Signal Post-Processing and Analysis Techniques Development Timeline

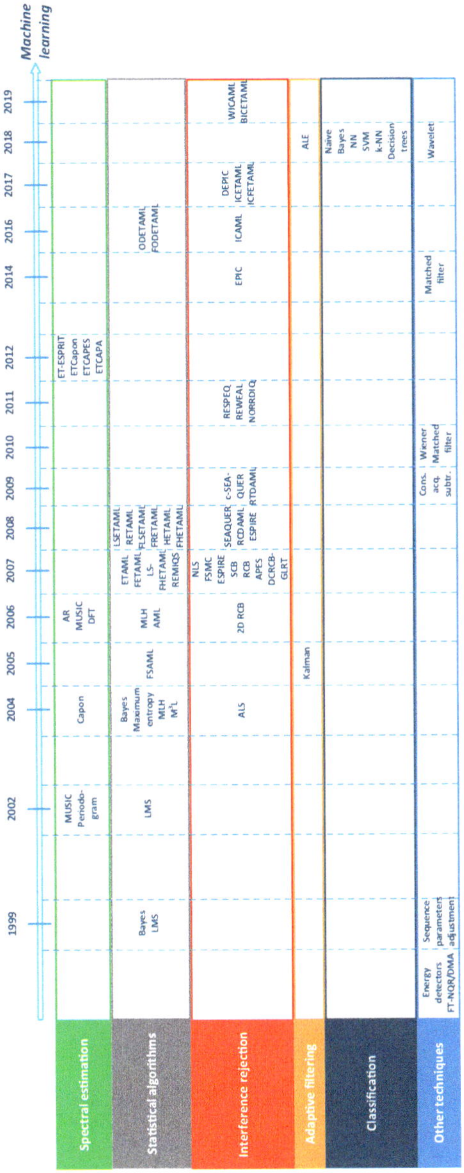
Machine learning
1999
2002
2004
2005
2006
2007
2008
2009
2010
2011
2012
2014
2016
2017
2018
2019
Spectral estimation
MUSIC Periodo-gram
Capon
AR MUSIC DFT
ET-ESPRIT ETCapon ETCAPES ETCAPA
Statistical algorithms
Bayes LMS
LMS
Bayes Maximum entropy MLH M²L
FSAML
MLH AML
ETAML FETAML LS-FHETAML REMIQS
LSETAML RETAML FLSETAML FRETAML HETAML FHETAML
ODETAML FODETAML
Interference rejection
ALS
2D RCB
NLS FSMC ESPIRE SCB RCB APES DCRCB-GLRT
SEAQUER RCDAML ESPIRE
c-SEA-QUER RTDAML
RESPEQ REWEAL NORRDIQ
EPIC
ICAML
DEPIC ICETAML ICFETAML
WICAML BICETAML
Adaptive filtering
Kalman
ALE
Classification
Naive Bayes NN SVM k-NN Decision trees
Other techniques
Energy detectors FT-NQR/DMA
Sequence parameters adjustment
Cons. acq. subtr.
Wiener Matched filter
Matched filter
Wavelet

Development of the Post-Processing Techniques

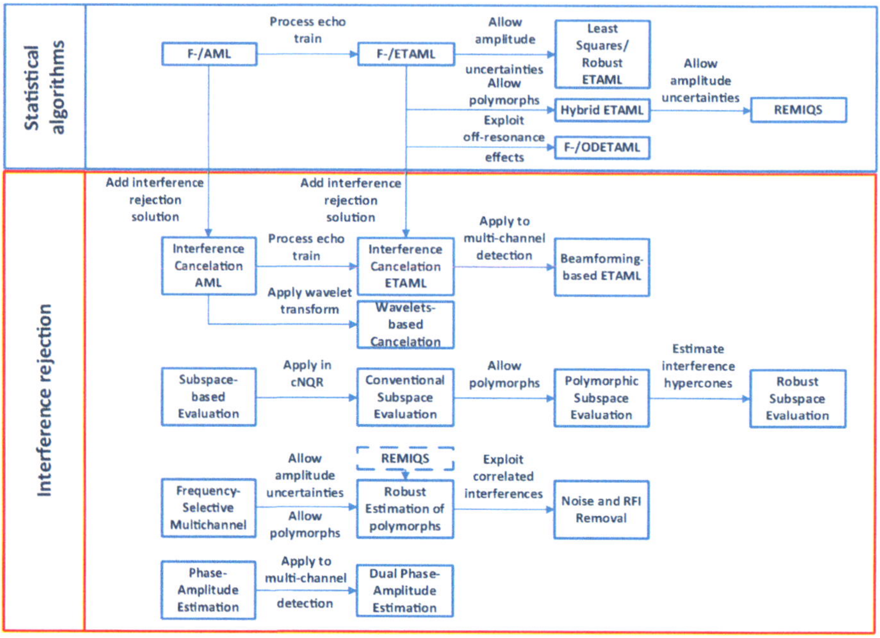

Distribution of the Post-Processing Techniques According to the Type of Detection [1][1]

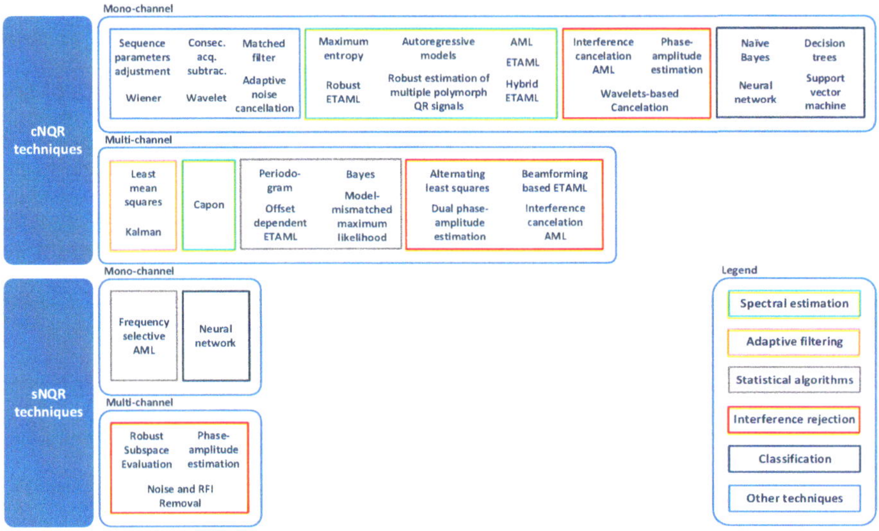

References

1. Monea, C.: A review of NQR signal processing and analysis techniques. J. Franklin Inst. **357**(17), 13085–13124 (2020)
2. Monea, C.: Signal processing and analysis methods in nuclear quadrupole resonance spectroscopy. J. Electrical Eng. Electron. Control Comput. Sci. **4**(2), 1–8 (2018)
3. Butt, N.R., Gudmundson, E., Jakobsson, A.: An Overview of NQR Signal Detection Algorithms, Magnetic Resonance Detection of Explosives and Illicit Materials, pp. 19–33 (2013)
4. Hemnani, P., Joshi, G., Rajarajan, A.K., Ravindranath, S.V.: 14N NQR spectrometer for explosive detection: a review. Int. Conf. Automat. Control Dyn. Optimizat. Tech. (2016)
5. Xinwang, Z.: A Low-power Compact Nuclear Quadrupole Resonance (NQR) Based Explosive Detection System. Ph.D thesis, University of Nebraska, Lincoln (2014)
6. Tagare, P.: Signal Averaging, Biomedical Digital Signal Processing, pp. 184–192. Prentice-Hall (1993)
7. NMR Analysis.: Sensitivity Enhancement for free? Internet: http://nmr-analysis.blogspot.com/2008/07/sensitivity-enhancement-for-free.html. 13 Aug. 2020
8. Kyriakidou, G.: Medicine Authentication using Nuclear Quadrulope Resonance. Ph.D. thesis, King's College London (2016)

[1] Reprinted from J. of the Franklin Inst., vol. 357, issue 17, C. Monea, A review of NQR signal processing and analysis techniques, Pages 13,085–13,124, Copyright 2020, with permission from Elsevier [OR APPLICABLE SOCIETY COPYRIGHT OWNER].

9. Vaseghi, S.V.: Advanced Digital Signal Processing and Noise Reduction, Fourth edn. Wiley (2008)
10. Somasundaram, S.D., Althoefer, K., Smith, A.S., Seneviratne, L.D.: Detection of landmines using nuclear quadrupole resonance (NQR): Signal processing to aid classification. In: Climbing and Walking Robots, pp. 833–840 (2006)
11. Kronval, S., Kronval, T.: Detection of Illegal Narcotics Using NQR. Masters thesis, University of Lund (2012)
12. Gudmundson, E., Wirfalt, P., Jakobsson, A., Jansson, M.: An ESPRIT-based parameter estimator for spectroscopic data. IEEE Stat. Signal Process. Workshop (2012)
13. Jakobsson, A., Mossberg, M., Rowe, M.D., Smith, J.A.S.: Exploiting temperature dependency in the detection of NQR signals. IEEE Trans. Signal Process. **54**(5), 1610–1616 (2006)
14. Jakobsson, A., Mossberg, M., Rowe, M.D., Smith, J.A.S.: Frequency-selective detection of nuclear quadrupole resonance signals. IEEE Trans. Geosci. Remote Sensing **43**(11), 2659–2665 (2005)
15. Somasundaram, S.D., Jakobsson, A., Gudmundson, E.: Exploiting spin echo decay in the detection of nuclear quadrupole resonance signals. IEEE Trans. Geosci. Remote Sensing **45**(4), 925–933 (2007)
16. Somasundaram, S.D., Jakobsson, A., Gudmundson, E.: Robust nuclear quadrupole resonance signal detection allowing for amplitude uncertainties. IEEE Trans. Signal Process. **56**(3), 887–894 (2008)
17. Somasundaram, S.D., Jakobsson, A., Smith, J.A.S.: Analysis of nuclear quadrupole resonance signals from mixtures. Signal Process. **88**(1), 146–157 (2008)
18. Butt, N., et al.: Robust detection of polymorphic NQR signals. In: 15th European Signal Processing Conference (EUSIPCO) (2007)
19. Somasundaram, S.D., et al.: Robust detection of stochastic nuclear quadrupole resonance signals. IEEE Trans. Signal Process. **56**(9), 4221–4229 (2008)
20. Somasundaram, S.D., et al.: Detecting stochastic nuclear quadrupole resonance signals in the presence of strong radio frequency interference. In: IEEE International Conference on Acoustics, Speech and Signal Process (2008)
21. Somasundaram, S.D., Jakobsson, A., Butt, N.R.: Countering radio frequency interference in single-sensor quadrupole resonance. IEEE Geosci. Remote Sensing Lett. **6**(1), 62–66 (2009)
22. Rudberg, T., Jakobsson, A.: Robust detection of nuclear quadrupole resonance signals in a non-shielded environment. In: Proceedings of the 19th European Signal Processing Conference (EUSIPCO) (2011)
23. Svensson, A., Jakobsson, A.: Adaptive detection of a partly known signal corrupted by strong interference. IEEE Signal Process. Lett. **18**(12), 729–732 (2011)
24. Swärd, J., Jakobsson, A.: Canceling stationary interference signals exploiting secondary data. In: 22nd European Signal Processing Conference (EUSIPCO) (2014)
25. Shao, W., Barras, J., Althoefer, K., Kosmas, P.: Detecting NQR signals severely polluted by interference. Signal Proces **138**, 256–264 (2017)
26. Shao, W., Kosmas, P., Althoefer, K., Barras, J.: Canceling strong and complex interference in NQR-based landmine detection. In: IEEE International Conference on Information and Automation for Sustainability (2016)
27. Shao, W., Barras, J., Kosmas, P.: A novel wavelets method for cancelling time-varying interference in NQR signal detection. Signal Process. **154**, 238–249 (2019)
28. Hemnani, P., Rajarajan, A.K., Joshi, G., Ravindranath, S.V.G.: The building of pulsed NQR/NMR sSpectrometer. Int. J. Electri. Comput. Eng. **8**(3), 1442–1450 (2018)
29. Hemnani, P., Rajarajan, A.K., Joshi, G., Ravindranath, S.V.G.: Detection of NQR signals using wavelet transform and adaptive filters. Int. J. Instrum. Tech. **2**(1), 34–49 (2018)
30. Cardona, L.R.: Nuclear Quadrupole Resonance System for Landmine Detection in Antioquia. Ph.D. thesis, National University of Colombia (2017)
31. Shao, W., Barras, J., Kosmas, P.: Detection of extremely weak NQR signals using stochastic resonance and neural network theories. Signal Process. **142**, 96–103 (2018)

32. Schiano, J., Routhier, T., Blauch, A.J., Ginsberg, M.D.: Feedback OPTIMIZATION OF PULSE WIdth in the SORC sequence. J. Magn. Reson. **140**, 84–90 (1999)
33. Schiano, J., Blauch, A.J., Ginsberg, M.D.: Optimization of NQR pulse parameters using feedback control, Z. Naturforsch **55**, 67–73 (2000)
34. Yang, T., et al.: NQR signal processing based on multi-stage wiener filter. Procedia Eng. **7**, 229–234 (2010)
35. Apostolos, J.T., Feng, J., Mouyos, W., McMahon, B.: Using phase matched filters for NQR detection of continuous Rabi transitions. Patent, US8660803B2 (2011)
36. Mozzhukhin, G.V., Kupriyanova, G.S., Mershiev, I.G., Molchanov, S.V.: Signal processing in NMR/NQR detection on the base of pattern signal. In: 9th International Kharkiv Symposium on Physics and Engineering of Microwaves, Millimeter and Submillimeter Waves (2016)
37. Yingyi, T., Tantum, S.L., Collins, L.M.: Landmine detection with nuclear quadrupole resonance. IEEE International Symposium on Geoscience and Remote Sensing (2002)
38. Jiang, Y., Stoica, P., Li, J.: Array signal processing in the known waveform and steering vector case. IEEE Trans. Signal Process. **52**(1), 23–35 (2004)
39. Tantum, S., et al.: Signal processing for NQR discrimination of buried landmines. In: SPIE Conference on Detection and Remediation Technologies for Mines and Minelike Targets IV, vol. 3710 (1999)
40. Stoica, P., Xiong, H., Xu, L., Li, J.: Adaptive beamforming for quadrupole resonance. Digital Signal Process. **17**(3), 634–651 (2007)
41. Xiong, H.: Robust Adaptive Methods and Their Applications in Quadrupole Resonance. Ph.D. thesis, University of Florida (2006)
42. Liu, G., Jiang, Y., Xiong, H., Li, J., Barrall, G.A.: Radio frequency interference suppression for landmine detection by quadrupole resonance. EURASIP J. Appl. Signal Process. **2006**, 1–14 (2006)
43. Xiong, H., Li, J., Barrall, G.A.: Joint TNT and RDX detection via quadrupole resonance. IEEE Trans. Aerospace Electron. Syst. **43**(4), 1282–1293 (2007)
44. Shao, W., Barras, J., Kosmas, P.: An advanced beamforming approach based on two-channel echo-train system to cancel interference within an NQR signal resonance spectrum. Signal Process. **154**, 136–147 (2019)
45. Jakobsson, A., Mossberg, M.: Using spatial diversity to detect narcotics and explosives using NQR signals. IEEE Trans. Signal Process. **55**(9), 4721–4726 (2007)
46. Butt, N.R., Jakobsson, A., Somasundaram, S.D., Smith, J.A.S.: Robust multichannel detection of mixtures using nuclear quadrupole resonance. IEEE Trans. Signal Process. **56**(10), 5042–5050 (2008)
47. Butt, N.R., Jakobsson, A.: Robust multi-sensor detection of polymorphic NQR signals. In: Conference on Record of the Forty-First Asilomar Conference on Signals, System and Computers (2007)
48. Butt, N.R., Jakobsson, A.: Efficient removal of noise and interference in multichannel quadrupole resonance. In: Conference on Record of the Forty Fifth Asilomar Conference on Signals, System and Computers (2011)
49. Piatti, T., Lei, S., Barras, J., Jakobsson, A.: Interference cancellation in two-channel nuclear quadrupole resonance measurements. IEEE Int. Conf. Acoustics Speech Signal Process. (2017)
50. Yingyi, T., Tantum, S.L., Collins, L.M.:, Kalman filtering for enhanced landmine detection using quadrupole resonance. IEEE Trans. Geosci. Remote Sensing **43**(7), 1507–1516 (2005)
51. ScienceDirect.: Căutare articole despre machine learning. Internet: https://www.sciencedirect.com/search?qs=machine%20learning&show=25&sortBy=relevance. 13 Aug. 2020
52. Póczos, B., Singh, A.L.: Introduction to machine learning CMU-10701 Deep Learning. Carnegie Mellon University. Internet: http://www.cs.cmu.edu/~aarti/Class/10701_Spring14/slides/DeepLearning.pdf. 4 Jan. 2019
53. Lazebnik, L.: Convolutional neural network architectures: from LeNet to ResNet. University of Illinois. Internet: http://slazebni.cs.illinois.edu/spring17/lec01_cnn_architectures.pdf. 13 Aug. 2020

54. Stfalcon.: Deep learning: definition, benefits, and challenges. Internet: https://stfalcon.com/en/blog/post/deep-learning-benefits-and-challenges. 4 Jan. 2019
55. Liu, X., Li, R., Zhao, C., Wang, P.: Robust signal recognition algorithm based on machine learning in heterogeneous networks. J. Syst. Eng. Electron. **27**(2), 333–342 (2016)
56. Klukowski, P., et al.: NMRNet: a deep learning approach to automated peak picking of protein NMR spectra. Bioinform. **34**(15), 2590–2597 (2018)

Chapter 5
Modeling of Signals Used in Nuclear Quadrupole Resonance Spectroscopy

The response signal generated by a substance analyzed using the NQR technique is characterized by several parameters [1]:

- Amplitude;
- Resonance frequency;
- Spectral linewidth;
- Quadrupolar coupling constant;
- Asymmetry parameter;
- Echo train decay time (T_{2e});
- Spin–spin relaxation time (T_2/T_2^*);
- Spin–lattice relaxation time (T_1).

These parameters are, in turn, dependent on the type of atoms, the molecular structure and the substance's temperature. Therefore, several signal models have been proposed to exploit these characteristics in order to increase the probability of detection.

5.1 Free Induction Decay Models

Initially, the FID signal was modeled using the following equation [2]:

$$y(t) = M_0 e^{\left(i\omega_k - T_2^{-1}\right)t} + w(t), 0 \leq t \leq T_{acq} \tag{5.1}$$

where M_0 represents the amplitude at time $=$ 0, ω_k is the resonance frequency, $T_{acq} \cong 4T_2$ is the acquisition time, and $w(t)$ is the random noise that is statistically independent from the response signal. This model may also take into account the noise associated with the piezoelectric and magnetoacoustic responses. The continuous time signal is illustrated in Fig. 5.1.

C. Monea and N. Bizon, *Signal Processing and Analysis Techniques for Nuclear Quadrupole Resonance Spectroscopy*, Signals and Communication Technology,
https://doi.org/10.1007/978-3-030-87861-0_5

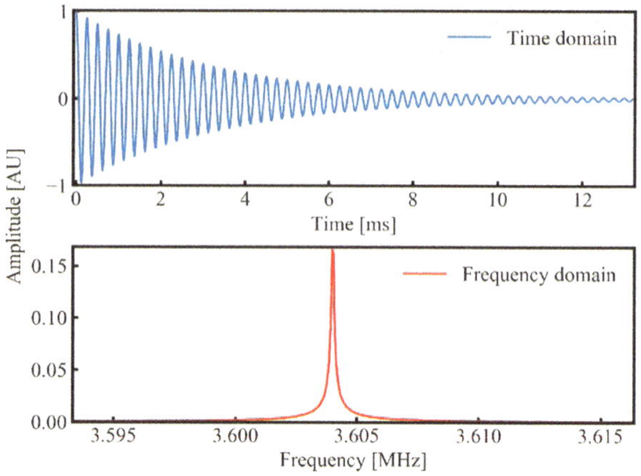

Fig. 5.1 FID signal modeled as a continuous time function, represented in the time (top) and frequency (bottom) domains (AU—arbitrary units)

The simulation is performed for sodium nitrite, in noise-free conditions, with the following parameters: $M_0 = 1$, $f_k = 3.604$ MHz, $T_2 = 3.3$ ms [3], and the sampling frequency is $f_s = 30$ MHz. The time domain signal represents the real part of the complex (FID) exponential, and it was decimated by a factor of 100 in order to observe it better.

The authors in [4–9] have proposed to model the FID signal as a sum of damped sinusoids, in noise, according to Eq. (4.5). Note that β_k, the damping constant associated with the kth sinusoid, depends on the spin–spin relaxation time (T_2^*) according to Equation [1]:

$$\beta_k = \frac{D_w}{T_2^*} \tag{5.2}$$

where D_w represents the sampling period [10]. Generally, all spectral lines have approximately the same damping constant, usually denoted by β_0, which does not vary significantly with the temperature [4]. This signal model has been used to develop algorithms applied for both cNQR and sNQR applications [4, 11]. Table 5.1 summarizes the values of the constants mentioned above, in the case of monoclinic TNT.

In Table 5.1, the spectral line amplitude of approximately 892 kHz is considered as a reference for determining the amplitudes for the other spectral lines. Figure 5.2

Table 5.1 The constants a_k, b_k, and K_k for 4 spectral lines of TNT [12]

Spectral line k	a_k (kHz)	b_k (kHz/K)	K_k
1	893.502	0.1529	0.2
2	875.734	0.1070	0.9
3	892.503	0.1685	1
4	870.293	0.1125	0.89

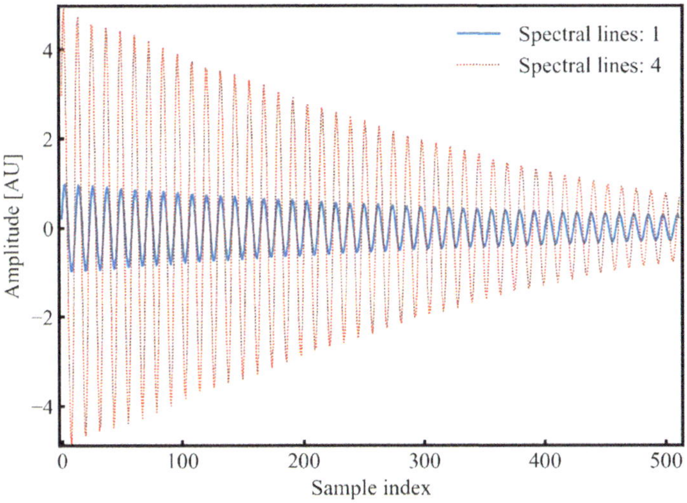

Fig. 5.2 FID signal modeled as a sum of damped sinusoids, composed of one and four spectral lines, respectively (AU—arbitrary units)

illustrates the real part of the FID signal composed of a single spectral line, and four resonance frequencies mentioned above, respectively, represented in 512 samples, with unitary ρ, $\beta_0 = 2.5 \cdot 10^{-4}$ and the sampling frequency $f_s = 10$ MHz. The noise is not included in order to simplify the plot, and the scaling constants K_k are considered real.

5.2 Echo Train Models

Similarly, the echo model is defined as two symmetric FID signals [4, 12], according to Eq. (4.7). Figure 4.3 shows the real part of an echo signal, without noise, with $t_0 = 512$ samples, $d = 1$ spectral line and $\beta_0 = 1.45 \cdot 10^{-2}$ to highlight the exponential decay. The sampling frequency is $f_s = 10$ MHz (Fig. 5.3).

This model has been extended to represent the entire echo train [5, 12–14], as presented in Eq. (4.10). The damping constant of the echo train corresponding to the kth spectral line, η_k, is given by the value of T_{2e}, according to the equality [1]:

$$\eta_k = \frac{D_w}{T_{2e}} \tag{5.3}$$

The amplitude and damping constants for the four spectral lines mentioned above were used to represent an echo train in the case of TNT. These are summarized in Table 5.2.

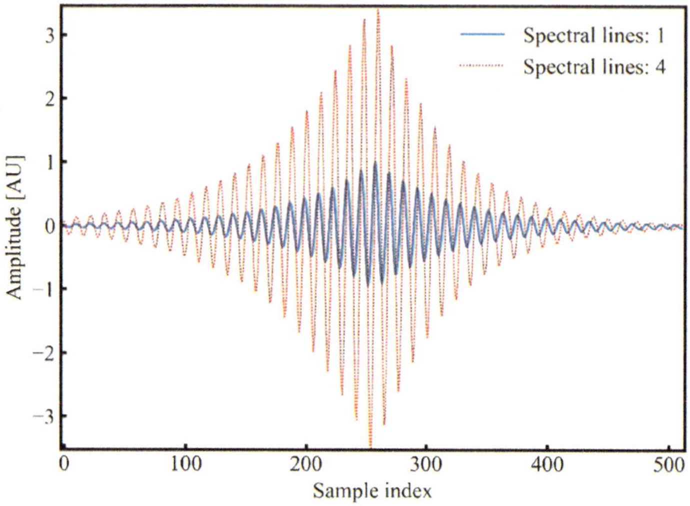

Fig. 5.3 Echo signal modeled as two symmetric FID signals, composed of one and four spectral lines, respectively (AU—arbitrary units)

Table 5.2 The constants β_k, η_k, and K_k for 4 spectral lines of TNT [5]

Spectral line k	$\beta_k \times 10^{-3}$	$\eta_k \times 10^{-3}$	$\lvert K_k \rvert$	$\angle K_k$ (°)
1	4.8	0.2126	0.39	−0.75
2	4.9	0.2096	0.88	−2.94
3	4.6	0.2237	1	−0.75
4	3.8	0.2576	0.69	−1.09

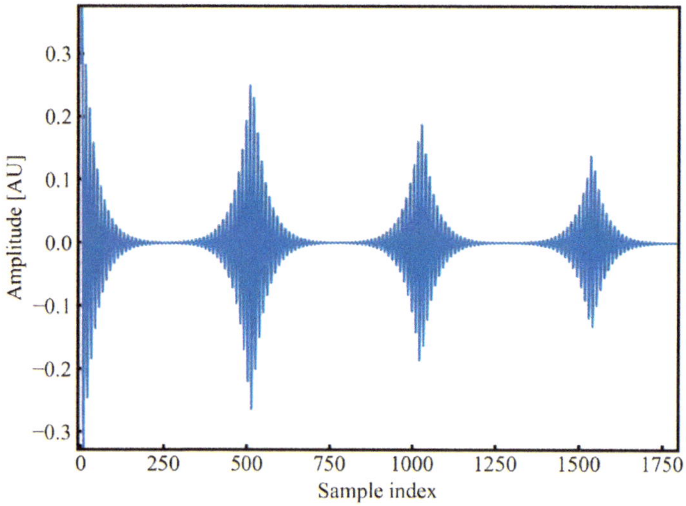

Fig. 5.4 The real part of an echo train (AU—arbitrary units)

Figure 5.4 shows the real part of an echo train generated with these parameters, composed of 3 echoes, with the length of 1800 samples, the space between echoes (measured against their centers) of 512 samples and containing a single resonance frequency. The sampling frequency is $f_s = 10$ MHz and β_k is multiplied by 5 to better highlight the shape of the echoes.

A similar echo train model is proposed in [15], where the mth echo is described by the equation:

$$y_m(t) = \sum_{k=1}^{d} \alpha_k \lambda_k^t \rho_k^m + w_m(t),\ \lambda_k = e^{-\beta_k - \eta_k + i\omega_k},\ \rho_k = e^{-2\tau\eta_k} \tag{5.4}$$

where τ represents half the length of an echo, and the other parameters are similar to those presented above. The parameters α_k, ω_k, β_k and η_k are approximated using spectral estimation algorithms such as Capon, CAPES and CAPA.

The echo train model has been extended to improve the probability of detection by evaluating the signals from several components of a mixture of polymorphic substances. Thus, the echoes are specific to each polymorphic form, according to the following equation [16, 17]:

$$y_m(t) = \sum_{k=1}^{p} \gamma_p y_m^{(p)}(t) + w_m(t) \tag{5.5}$$

where $y_m^{(p)}$ is defined in Ec. (4.7), but has the model parameters specific to the polymorphic form p, and γ_p represents the proportion of the p component in the mixture.

A different NQR signal model takes into account the off-resonance effects, which represent mismatches between the excitation and the resonance frequency, caused by the temperature gradient within the substance [1, 18].

The FID signal model presented in Eq. (4.5) is used in the sNQR interference rejection algorithms [19, 20]. If the substance is analyzed using P stochastic pulses, and after each of them N samples are acquired, then the time domain signal will contain NP samples. Its correlation with the excitation sequence will provide the signal $r(t)$, having the same number of samples, which can be modeled as an FID signal consisting of equally spaced sample blocks, with the spaces between them depending on the sampling time. Thus, the first block of samples is expressed as follows:

$$r^p(t) = \sum_{k=1}^{d} \alpha_k e^{[-\beta_k + i\omega_k(T)](t + pT_s)} + w^p(t),\ p = 0, \ldots, P-1 \tag{5.6}$$

where $t = t_0, \ldots, t_{N-1}$, T_s, d and T represent the sampling time of a data block (measured from the center of a stochastic pulse), the time between two excitation pulses, the known number of spectral lines and the unknown temperature of the substance, respectively.

5.3 Proposal of a New Echo Model

Although it is used in the development of NQR detection algorithms, the echo signal model presented in Eq. (4.10) is not found in theory [21–25], nor in practical experiments [26–31]. Specifically, the sharp shape of the echoes does not resemble the response of a substance excited by the SLSE technique. For the most accurate simulation of the NQR response, an echo model obtained by applying the Kaiser window to the model presented in Eq. (4.7) is proposed. The Kaiser window is mainly used in the design of numerical filters with finite impulse response and in spectral analysis, and was chosen because its shape approximates that of the echo. The shape of the window is given by its size N and the shape parameter β [32]. Thus, the echo train can be expressed as follows:

$$y_m(t) = \sum_{k=1}^{d} \alpha_k e^{\left(-\eta_k\left(t+2mt_{sp}\right)-\beta_k k_s(t/T)+i\omega_k(T)t\right)} + w_m(t) \tag{5.7}$$

where k_s represents the window's coefficients of size equal to the number of samples corresponding to an echo, and T is the sampling period. Figure 5.5 illustrates the Kaiser window generated with the parameters $N = 2560$ samples and $\beta = 10$, and

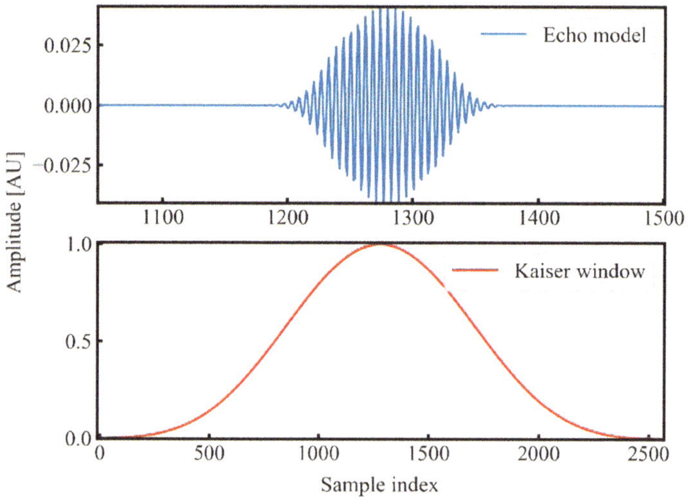

Fig. 5.5 Proposed echo signal (top) and the Kaiser window applied (bottom) (AU—arbitrary units)

the proposed model obtained by multiplying the echo in Eq. (4.7) by the window's coefficients, respectively.

A comparative study of several signal models for more accurate echo representation was performed in [33], and the Pearson type VII model provided the best results.

5.4 Noise Models

One type of noise that is often used in the study of algorithms applied in NQR spectroscopy is the white and additive Gaussian noise [20, 34, 35]. It is characterized by a power spectral density that is uniformly distributed throughout the investigated band, the lack of correlation between any two samples and a normal distribution described by the probability density function:

$$f(x) = \frac{1}{\sqrt{2\pi\sigma^2}} e^{-\frac{(x-\mu)^2}{2\sigma^2}} \tag{5.8}$$

where μ represents the average value, and σ^2 is the variance. This function is customized for a one-dimensional random variable and can be extended for a multidimensional case. Colored Gaussian additive noise is also applied in algorithm simulations [36, 37] and differs from white noise by its power spectral density characteristics.

Often, the noise is represented by a small-order autoregressive model estimated from the acquired data [5, 12, 16]. It is defined by the equation:

$$w(n) = c + \sum_{i=0}^{p-1} c_i w(n-i) + \varepsilon(n) \tag{5.9}$$

where c represents a constant, c_i are the model's parameters and $\varepsilon(n)$ is white noise.

Other studies have used non-Gaussian colored noise, modeled by an autoregressive model [7, 38]. Radio frequency interference is also modeled as the sum of complex sinusoids, as follows:

$$r(n) = \sum_{i=1}^{M} A_i(n) e^{j(2\pi f_i n + \phi_i)}, n = 0, \ldots, N- \tag{5.10}$$

where $A_i(n)$ represents the complex amplitude, which can be, e.g., the signal emitted by a commercial radio station, and f_i and ϕ_i are the frequency and phase of each sinusoidal component, respectively. The number of samples is N and the number of sinusoids is M. A simplified case is simulated in [13], where 9 stationary components are considered, with constant (real) amplitude, random phases and frequencies defined in relation to the sampling frequency.

5.5 Conclusions

This chapter has presented a review of the main signal models applied in nuclear quadrupole resonance spectroscopy. Several free induction decay, noise, interference, echo and echo train models were described and simulated. A new spin echo model was proposed based on the Kaiser window in order to simulate more accurately its real shape. The models' implementation and the results presented in this chapter are used for carrying out a study of the signal processing and analysis algorithms applied in NQR detection.

References

1. Kyriakidou, G.: Medicine Authentication Using Nuclear Quadrulope Resonance. Ph.D. thesis, King's College London (2016)
2. Gerothanassis, I.P.: Methods of avoiding t[18]he effects of acoustic ringing in pulsed fourier transform nuclear magnetic resonance spectroscopy. Prog. Nucl. Magn. Reson. Spectroscopy **19**(3), 267–329 (1987)
3. Mandal, S., Song, Y.Q.: Method and system for applying NQR pulse sequences. Patent, WO2013134474 (2013.
4. Somasundaram, S.D., Althoefer, K., Smith, A.S., Seneviratne, L.D.: Detection of landmines using nuclear quadrupole resonance (NQR): signal processing to aid classification. In: Climbing and Walking Robots, pp. 833–840 (2006)
5. Somasundaram, S.D., Jakobsson, A., Gudmundson, E.: Exploiting spin echo decay in the detection of nuclear quadrupole resonance signals. IEEE Trans. Geosci. Remote Sensing **45**(4), 925–933 (2007)
6. Lawes, D.J.M., Mallion, S., Rowe, M.D., Smith, J.A.S.: Time-domain data analysis of NQR response. Patent, 9915842.0 (1999)
7. Tan, Y., Tantum, S.L., Collins, L.M.: Cramer-Rao lower bound for estimating quadrupole resonance signals in non-gaussian noise. IEEE Signal Process. Lett. **11**(5), 490–493 (2004)
8. Smith, J.A.S., Peirson, N.F.: Method and apparatus for NQR testing a sample and pulse sequence for exciting NQR. Patent, WO9626453 (1996)
9. Lawes, D.J.M., Mallion, S., Rowe, M.D., Smith, J.A.S.: NQR signal processing. Patent, GBOO/02 582 (2000)
10. Gudmundson, E., Stoica, P., Li, J., Jakobsson, A., Rowe, M.D., Smith, J.A.S., Ling, J.: Spectral estimation of irregularly sampled exponentially decaying signals with applications to RF spectroscopy. J. Magn. Reson. **203**(1), 167–176 (2010)
11. Gudmundson, E.: Signal Processing for Spectroscopic Applications. Ph.D. thesis, Uppsala University, Sweden (2010)
12. Jakobsson, A., Mossberg, M., Rowe, M.D., Smith, J.A.S.: Exploiting temperature dependency in the detection of NQR signals. IEEE Trans. Signal Process. **54**(5), 1610–1616 (2006)

13. Shao, W., Barras, J., Althoefer, K., Kosmas, P.: Detecting NQR signals severely polluted by interference. Signal Proces. **138**, 256–264 (2017)
14. Butt, N.R., Gudmundson, E., Jakobsson, A.: An overview of NQR signal detection algorithms. In: Magnetic Resonance Detection of Explosives and Illicit Materials, pp. 19–33 (2013)
15. Kronval, S., Kronval, T.: Detection of Illegal Narcotics Using NQR. Masters thesis, University of Lund (2012)
16. Somasundaram, S.D., Jakobsson, A., Smith, J.A.S.: Analysis of nuclear quadrupole resonance signals from mixtures. Signal Process. **88**(1), 146–157 (2008)
17. Butt, N.R., Somasundaram, S.D., Jakobsson, A., Smith, J.A.S.: Frequency-selective robust detection and estimation of polymorphic QR signals. Signal Process. **88**(4), 834–843 (2008)
18. Kyriakidou, G., Jakobsson, A., Gudmundson, E., Gregorovič, AT., Barras, J., Althoefer, K.: Improved modeling and bounds for NQR spectroscopy signals. In: 22nd European Signal Processing Conference, pp. 2325–2329 (2014)
19. Somasundaram, S.D., et al.: Robust detection of stochastic nuclear quadrupole resonance signals. IEEE Trans. Signal Process. **56**(9), 4221–4229 (2008)
20. Somasundaram, S.D., et al.: Detecting stochastic nuclear quadrupole resonance signals in the presence of strong radio frequency interference. In: IEEE International Conference on Acoustics, Speech and Signal Processing (2008)
21. Coy, A., Callaghan, P.T.: Pulsed gradient spin echo nuclear magnetic resonance for molecules diffusing between partially reflecting rectangular barriers. The J. Chem. Phys. **101**(6), 4599–4609 (1994)
22. Hacking, C., Ray Ballinger, J., et al.: Spin echo sequences, Radiopaedia. Internet: https://radiopaedia.org/articles/spin-echo-sequences. 18 Aug. 2020.
23. Bottomley, P.: Selective volume method for performing localized NMR spectroscopy. Patent US4480228A (1984)
24. Questions and Answers in MRI, Fast spin-echo. Internet: http://mriquestions.com/what-is-fsetse.html. 18 Aug. 2020
25. Saito, H., Kato, S., Matsumoto, K., Umino, Y., Kita, R., Shinyashiki, N., Yagihara, S., Fukuzaki, M., Tokita, M.: Dynamic behaviors of solvent molecules restricted in poly (acryl amide) gels analyzed by dielectric and diffusion NMR spectroscopy. Gels **4**(3), 56 (2018)
26. Hahn, E.L.: Spin echoes. Phys. Rev. **80**(4), 580–594 (1950)
27. Stoltenberg, J., Pengra, D., Vilches, O., Dyck, R.V.: Pulsed Nuclear Magnetic Resonance. Internet: http://instructor.physics.lsa.umich.edu/adv-labs/NMR/pulsed_NMR_rev.pdf. 18 Aug. 2020
28. Stallmach, F., Galvosas, P.: Spin echo nmr diffusion studies. In: Annual Report on NMR Spectroscopy, pp. 51–131 (2007)
29. Freedman, R., Anand, V., Grant, B., Ganesan, K., Tabrizi, P., Torres, R., Catina, D., Ryan, D., Borman, C., Krueckl, C.: A compact high-performance low-field NMR apparatus for measurements on fluids at very high pressures and temperatures. Rev. Sci. Instrum. **85**(2), 025102 (2014)
30. Beguš, S., Jazbinšek, V., Pirnat, J., Trontelj, Z.: A miniaturized NQR spectrometer for a multichannel NQR-based detection device. J. Magn. Reson. **247**, 22–30 (2014)
31. Christian, G., Hermann, S.: NQRS raw data to https://doi.org/10.1039/C8CP03848A TABLE 1, Zenodo. Internet: https://zenodo.org/record/1462856#.XzvpzWziuM8. 18 Aug. 2020
32. Oppenheim, A.V., Schafer, R.W.: Discrete-Time Signal Processing, 3rd edn. Pearson (2009)
33. Monea, C.: Improved signal models for nuclear quadrupole resonance spectroscopy. In: 12th International Conference on Electronics, Computers and Artificial Intelligence (2020)
34. Shao, W., Barras, J., Kosmas, P.: A novel wavelets method for cancelling time-varying interference in NQR signal detection. Signal Process. **154**, 238–249 (2019)
35. Yingyi, T., Tantum, S.L., Collins, L.M.: Landmine detection with nuclear quadrupole resonance. IEEE Int. Geosci. Remote Sensing Symp. (2002)
36. Butt, N.R., Jakobsson, A., Somasundaram, S.D., Smith, J.A.S.: Robust multichannel detection of mixtures using nuclear quadrupole resonance. IEEE Trans. Signal Process. **56**(10), 5042–5050 (2008)

37. Butt, N.R., Jakobsson, A.: Efficient removal of noise and interference in multichannel quadrupole resonance. In: Conference on Record of the Forty Fifth Asilomar Conference on Signals, Syst. and Computers (2011)
38. Yingyi, T., Tantum, S.L., Collins, L.M.: Kalman filtering for enhanced landmine detection using quadrupole resonance. IEEE Trans. Geosci. Remote Sensing. **43**(7), 1507–1516 (2005)

Chapter 6
Study of the NQR Signal Processing and Analysis Algorithms

Currently, there are dozens of solutions developed for NQR, and this study selects only several significant solutions in order to make a comparative analysis. As presented in a previous chapter, the most recent solutions are those specific machine learning. Therefore, it is justified to choose classification solutions for the comparison. Thus, the following classification algorithms are taken into consideration: k-NN, SVM, NN and EC.

At the same time, the statistical solutions represent a large part of the existing solutions and are used for developing other techniques. Thus, RETAML is considered, as well as its similar counterpart, LSETAML. An algorithm that can exploit the responses from polymorphic substances or off-resonance effects was not chosen, because it is designed for particular detection situations. In addition, only cNQR solutions are considered, because sNQR ones are less utilized.

In order to give a higher degree of novelty to this comparative analysis, a multi-criteria algorithm for detection is proposed, presented in three optimized variants.

6.1 Description of the Algorithms Chosen for the Analysis

The classification algorithms were used in [1, 2] to detect ammonium nitrate landmines. These solutions aim to identify the response based on the following features:

- C_1—amplitude of the spectral component at the reference frequency (423 kHz);
- C_2—the sum of the amplitudes of the spectral components in the band 422.5–423.8 kHz;
- C_3—the sum of the amplitudes of the spectral components in the band 420–426.3 kHz;
- C_4—the frequency of the highest spectral peak in the band 400–450 kHz;

C. Monea and N. Bizon, *Signal Processing and Analysis Techniques for Nuclear Quadrupole Resonance Spectroscopy*, Signals and Communication Technology,
https://doi.org/10.1007/978-3-030-87861-0_6

- C_5—spectral component's frequency that is closest to the reference;
- C_6—spectral component's amplitude that is located at frequency C_5;
- C_7—ratio between the spectral component's amplitude that is located 1.5 kHz left of C_5, and C_6;
- C_8—ratio between the spectral component's amplitude that is located 1.5 kHz right of C_5, and C_6;
- C_9—C_1/S_{ref};
- C_{10}—C_2/S_{ref};
- C_{11}—C_3/S_{ref},

where S_{ref} represents the spectral average in the bands 400–420 kHz and 426.5–450 kHz.

Several features are used to increase the probability of detection, allowing the correct identification of the response in several scan scenarios. Since NQR detection is a binary classification problem, two signal classes were used: the "mine" class, containing the data sets that contain the NQR signal, and the "other" class, containing the data sets without the response signal.

The **k-NN** algorithm searches each data set for the first k closest neighbors in the training set. Then it is associated with the most common class to these neighbors. If $k = 1$, then it is associated with the class of the nearest neighbor. 7 neighbors were used, and the best performance was obtained using the first 5 most significant features indicated by principal component analysis.

The **SVM** classifier separates the classes using a hyperplane. The purpose of this algorithm is to orient this hyperplane as far away from the members of both classes that are located closest to it. Using all the features ensured the best detection performance.

NN is a classifier inspired by the human brain and is composed of processing elements called artificial neurons. The output of each neuron is a function of the weighted sum of the inputs. A network with 11 input, 5 hidden and 1 output neurons was used, and the best performance was obtained using all the features.

In the case of **EC**, the decisions of multiple classifiers are combined (usually by weighting). The Bagging method was used to assemble the classifiers, forming a collection of 20 decision trees, and the best performance was obtained using all the features.

The **RETAML** algorithm is an improved version of ETAML that estimates the vector of complex amplitudes, using previously known information about it. The data model utilized is based on the one described by Eq. (4.10), and is defined as follows:

$$z^m(t) = \sum_{k=1}^{d} \bar{C}\alpha_k e^{-\eta_k(t+2mt_{sp})-\beta_k|t-t_{sp}|+i\omega_k(T)t} + e^m(t) \tag{6.1}$$

where $t = t_0, \ldots, t_{N-1}$ is the sample number, $m = 0, \ldots, M-1$ is the echo number, $e^m(t)$ represents the additive white noise, and $\overline{C}$ is defined as follows:

$$\overline{C} = \begin{cases} C(\lambda_k), \text{ for } t - t_{sp} < 0 \\ C\left(\tilde{\lambda}_k\right), \text{ in rest} \end{cases} \tag{6.2}$$

where x is the integer part of x, $\lambda_k = e^{i\omega_k(T)+\beta_k-\eta_k}$, $\tilde{\lambda}_k = e^{i\omega_k(T)-\beta_k-\eta_k}$, and $C(\lambda_k)$ is used for pre-whitening. Based on this mathematical formulation, the mth echo can be described as follows:

$$z_N^m = \left[z^m(t_0)\ldots z^m(t_{N-1})\right]^T = \left(\boldsymbol{A}_{\overline{\boldsymbol{\theta}}} \odot \boldsymbol{C}\right)\boldsymbol{B}_m\alpha + \boldsymbol{e}_N^m \tag{6.3}$$

where $\boldsymbol{e}_N^m$ is the noise associated with the mth echo, defined similarly to z_N^m, $[\cdot]^T$ is the transposed vector/matrix, and $\odot$ represents the Schur-Hadamard product. The signal model's components are defined as follows:

$$\boldsymbol{A}_{\overline{\boldsymbol{\theta}}} = \begin{bmatrix} \zeta_1^{t_0} e^{-\beta_1|t_0-t_{sp}|} & \ldots & \zeta_d^{t_0} e^{-\beta_d|t_0-t_{sp}|} \\ \vdots & \ddots & \vdots \\ \zeta_1^{t_{N-1}} e^{-\beta_1|t_{N-1}-t_{sp}|} & \ldots & \zeta_d^{t_{N-1}} e^{-\beta_d|t_{N-1}-t_{sp}|} \end{bmatrix} \tag{6.4}$$

$$C = \begin{bmatrix} C(\lambda_1) & \cdots & C(\lambda_d) \\ \vdots & \ddots & \vdots \\ C(\lambda_1) & \cdots & C(\lambda_d) \\ C\left(\tilde{\lambda}_1\right) & \cdots & C\left(\tilde{\lambda}_d\right) \\ \vdots & \ddots & \vdots \\ C\left(\tilde{\lambda}_1\right) & \cdots & C\left(\tilde{\lambda}_d\right) \end{bmatrix} \tag{6.5}$$

$$\boldsymbol{B}_m = diag\left\{\left[e^{-\eta_1 2mt_{sp}}\ldots e^{-\eta_d 2mt_{sp}}\right]\right\} = \begin{bmatrix} e^{-\eta_1 2mt_{sp}} & \ldots & 0 \\ \vdots & \ddots & \vdots \\ 0 & \ldots & e^{-\eta_d 2mt_{sp}} \end{bmatrix} \tag{6.6}$$

$$\alpha = [\alpha_1 \ldots \alpha_d]^T \tag{6.7}$$

where $\zeta_k = e^{i\omega_k(T)-\eta_k}$, and $\overline{\boldsymbol{\theta}} = \left[T\boldsymbol{\beta}^T\boldsymbol{\eta}^T\right]$ is the vector of unknown variables to be estimated (the substance's temperature, the damping constants vector for the d sinusoids and for the echo trains related to the d spectral lines, respectively).

Using these notations, the echo train is described mathematically as follows:

$$z_{NM} = \left[\left(z_N^0\right)^T\ldots\left(z_N^{M-1}\right)^T\right]^T = \boldsymbol{H}_{\overline{\theta}}\alpha + \boldsymbol{e}_{NM} \tag{6.8}$$

where $\boldsymbol{e}_{NM}$ is defined similarly to z_{NM}, and

$$H_{\overline{\theta}} = \begin{bmatrix} \left(A_{\overline{\theta}} \odot C\right) B_0 \\ \vdots \\ \left(A_{\overline{\theta}} \odot C\right) B_{M-1} \end{bmatrix} \tag{6.9}$$

The RETAML algorithm assumes that $\boldsymbol{\alpha}$ is proportional to the amplitude vector corresponding to the d spectral lines $\boldsymbol{\kappa}$, as follows: $\boldsymbol{\alpha} = \rho\boldsymbol{\kappa}$, where ρ is the scaling constant with the signal strength. Then, RETAML estimates the amplitude vector in an uncertainty sphere with radius $\sqrt{\varepsilon}$, starting from an assumed value $\overline{\boldsymbol{\kappa}}$, by performing the following constrained minimization:

$$\min_{\kappa} \left\| \rho H_{\bar{\theta}} \kappa - z_{NM} \right\|_2^2, \; cu \; \|\kappa - \bar{\kappa}\|_2^2 \leq \varepsilon \tag{6.10}$$

where $\|\cdot\|_2$ is the Euclidean norm. The estimated vector $\hat{\boldsymbol{\kappa}}$ is used to estimate $\boldsymbol{\alpha}$, thus obtaining $\hat{\boldsymbol{\alpha}}$. Finally, the following statistical test is performed:

$$\mathrm{T}\left(z_{NM}, \hat{\alpha}\right) = \frac{\|z_{NM}\|_2^2}{\left\|z_{NM} - H_{\bar{\theta}}\hat{\alpha}\right\|_2^2} \tag{6.11}$$

and considering the substance detected if $T\left(z_{NM}, \hat{\boldsymbol{\alpha}}\right) > \gamma$, where γ is a detection threshold associated with the accepted probability of false alarms.

The **LSETAML** algorithm estimates the unknown amplitude vector $\boldsymbol{\alpha}$ in order to recreate (estimate) the echo train, z_{NM}. It is similar to RETAML, with the difference that the amplitude vector is estimated by using the least squares (LS) method. As in RETAML, a search region is established for the values of the vector $\overline{\boldsymbol{\theta}}$, finding those that maximize a cost function, thus resulting the estimated $\hat{\overline{\boldsymbol{\theta}}}$ defined as follows:

$$\hat{\overline{\theta}} = \arg \max_{\overline{\theta}} z_{NM}^{*} H_{\overline{\theta}} H_{\overline{\theta}}^{\dagger} z_{NM} \tag{6.12}$$

where $(\cdot)^{*}$ is the conjugate transpose, and $(\cdot)^{\dagger}$ is the Moore–Penrose pseudoinverse. Finally, the following statistical test is performed:

$$\mathrm{T}(z_{NM}) = (2NM - 1)\frac{z_{NM}^{*} H_{\overline{\theta}} H_{\overline{\theta}}^{\dagger} z_{NM}}{z_{NM}^{*}\left(I - H_{\overline{\theta}} H_{\overline{\theta}}^{\dagger}\right) z_{NM}} \tag{6.13}$$

where the resulting value is compared with the threshold γ.

6.2 Proposal of a Multi-criteria Detection Algorithm

In order to offer a higher degree of novelty to this comparative analysis, an algorithm based on several criteria is proposed to improve the probability of detection. This solution is a different approach from the existing solutions because it does not require estimating the statistical parameters of the NQR response and is not dependent on a signal model. The solution uses several algorithms based on Fourier spectral analysis, which provide advantages in specific detection scenarios, and performs the detection decision by evaluating the weighted average of their results. In the following, the solution will be called multi-criteria Fourier analysis decision algorithm (MFD).

The MFD algorithm represents the fusion of six criteria (algorithms) based on statistical values (statistical detectors) and the maximum energy associated with some frequencies within the detection band (energy detectors) [3]. The statistical criteria are added to those based on signal energy to increase the probability of detection in the case of signal acquisitions with low SNR, when the latter cannot achieve good performance [150]. The detection decision is made by evaluating the weighted average of the results. Signal processing is performed in the frequency domain, because the response can be seen more easily than in the time domain; thus, all the algorithms are based on the complex discrete Fourier transform calculated using the fast Fourier transform (FFT) algorithm.

The first algorithm initially assumes that the substance is detected and determines the amplitude of the maximum spectral peak. Then, it checks for another spectral peak in the search band around the center frequency, which exceeds an amplitude threshold. Considering the received signal vector $x[n]$ of length N, where n represents the sample index, then the algorithm decides that the substance has not been detected if:

$$X[k] \geq A, k = \{i : -B \leq F(i) \leq B, i \neq k_{max}, X[i] \text{ is peak}\} \tag{6.14}$$

where k represents the set of search indices,

$$X[k] = \text{DFT}\{x[n]\} \tag{6.15}$$

A is the amplitude threshold of the spectral peak defined as a percentage A_p of the maximum spectral peak's amplitude,

$$A = A_p \max(\boldsymbol{X}) \tag{6.16}$$

B is the search frequency band around the center frequency, that is also expressed as a percentage B_p of the spectral bandwidth (S_W) used to acquire the signal,

$$B = B_p S_W \tag{6.17}$$

k_{max} is the spectral peak index with the highest amplitude, and the function $F(k)$ returns the frequency of the kth spectral component,

$$F(k) = kS_W/N_{DFT} \tag{6.18}$$

where N_{DFT} represents the DFT's dimension.

As observed in Eq. (6.14), the algorithm checks all peaks in the investigated frequency band and if one (other than the maximum peak) is identified with an amplitude at least equal to the threshold value, then it decides that the NQR response is absent. In this case, the spectrum contains peaks with amplitudes comparable to that of the maximum peak, and the criterion interprets it as a spectrum that contains only noise. This algorithm should allow the identification of the NQR response in the presence of RFI that does not spectrally overlap the response.

The second algorithm is similar to the first, with the difference that the spectral peak search band is centered on the maximum peak's frequency and not on the center frequency. It decides that the substance has not been detected if:

$$X[k] \geq A, k = \left\{i : -B' \leq F(i) \leq B', i \neq k_{\max}, X[i] \text{ is peak}\right\} \tag{6.19}$$

where

$$B' = B + F(k_{\max}) \tag{6.20}$$

This criterion should allow the identification of the NQR response in scenarios with low SNR and frequency deviation (when the spectral peak can be shifted from the excitation frequency).

The third algorithm is different from the first two in that it searches the entire spectrum. It decides that the substance has not been detected if:

$$X[k] \geq A, k = \{i : -S_W \leq F(i) \leq S_W, i \neq k_{max}, X[i] \text{ is peak}\} \tag{6.21}$$

This criterion should allow the detection of NQR response in scenarios with low SNR. Typical values for A_p and B_p are 0.7 and 0.785, respectively, and S_W varies from 50 to 200 kHz.

The fourth algorithm compares the spectral average with an amplitude threshold defined as a percentage of the maximum spectral peak. It decides that the substance has not been detected if:

$$\overline{X} = (1/N_{DFT}) \sum_{k=0}^{N_{DFT}-1} X[k] \geq A \tag{6.22}$$

In this case, a typical value for A_p is 0.214. This algorithm should allow the identification of acquisitions that contain only noise, when the average value is comparable to that of the maximum spectral peak. If a peak response stands out from the noise, then this criterion decides that the substance is detected.

The purpose of the first four criteria is to determine whether the received signal's spectrum contains the useful signal (NQR response) and noise or only noise, in the case of acquisitions where the substance response is weak and surrounded by noise-associated spectral peaks.

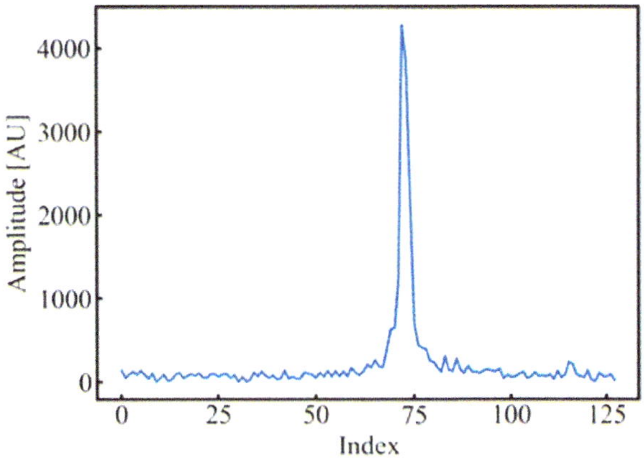

Fig. 6.1 Reference spectrum (AU—arbitrary units)

The fifth algorithm processes the received signal by correlating it with a reference spectrum, in order to identify the response peak. The reference spectrum is illustrated in Fig. 6.1.

The reference spectrum is obtained by scanning a sample of sodium nitrite in optimal conditions (high SNR) and has a minimum noise level and a clear spectral peak. Since all substances respond similarly in good conditions, this spectrum can be used as a reference for comparison.

This algorithm first normalizes the two spectra. Then, it compares their lengths and equalizes them by adding zeros at both ends, if necessary. At this stage, both vectors have the same length, M. After that, it aligns the response peaks by shifting the reference spectrum by a value determined by the cross-correlation of the two spectra. Finally, the algorithm calculates the Pearson correlation coefficient [4] and compares it with a threshold. It decides that the substance has been detected if:

$$\mathrm{PCC}(\boldsymbol{X}_n, \boldsymbol{R}_n) = \frac{\sum_{k=0}^{M-1}\left(X_n[k] - \overline{X_n}\right)\left(R_n[k] - \overline{R_n}\right)}{\sqrt{\sum_{k=0}^{M-1}\left(X_n[k] - \overline{X_n}\right)^2}\sqrt{\sum_{k=0}^{M-1}\left(R_n[k] - \overline{R_n}\right)^2}} \geq C_t \tag{6.23}$$

where $\overline{X_n}$ and $\overline{R_n}$ are the average values of the normed spectra of the response and the reference signals, respectively, defined similarly to Eq. (6.22), $\boldsymbol{X}_n$ and $\boldsymbol{R}_n$ are the normed vectors,

$$X_n[k] = X[k]/\max(\boldsymbol{X}) \tag{6.24}$$

$$R_n[k] = R[k]/\max(\boldsymbol{R}) \tag{6.25}$$

and C_t is the correlation threshold. Usually, C_t is equal to 0.57.

The lengths of the two spectra may differ, because the length of the response spectrum depends on the spectral width used to acquire the signal, which varies from one substance to another. This is due to the fact that the substances that respond more strongly to the excitation allow the use of a high spectral bandwidth, while those with a weaker response must be detected using a lower spectral bandwidth in order to acquire less noise. This criterion should allow the detection of the NQR response in scenarios with very low SNR.

The sixth algorithm compares the average values of the band around the response peak's frequency and the side bands. Thus, the algorithm would be useful in very noisy acquisitions, where the response peak cannot be visually identified within the spectrum, or in cases where the spectral response line is broadened due to the non-uniform substance temperature. The center band has a fixed width and is positioned on the maximum peak; its average value is defined as follows:

$$\overline{X_c} = (1/N_c)\sum_{k\in C} X[k],\ C = \{i : -[F(k_{\max}) - B_c] \le F(i) \le F(k_{\max}) + B_c\} \tag{6.26}$$

where C is the set of indices associated with the frequencies within the center band, N_c is the number of elements contained in C, and B_c is the bandwidth, usually equal to 3 kHz. Similarly, the average value of the sidebands is defined as follows:

$$\overline{X_s} = (1/N_s)\sum_{k\in S} X[k],\ S = \{i : F(i)\langle -[F(k_{\max}) - B_c],\ F(i)\rangle F(k_{\max}) + B_c\} \tag{6.27}$$

where S is the set of indices associated with frequencies outside the center band and N_s is the number of elements contained in S. If the peak is identified within a 10 kHz band centered on the reference frequency, then the algorithm considers it an NQR response peak and C centers on it. Otherwise, C focuses on the reference frequency. This special condition avoids the situation where a noise peak is considered an NQR response.

The algorithm decides that the substance has been detected if:

$$\overline{X_c}/\overline{X_s} \ge M_t \tag{6.28}$$

where M_t represents the threshold for the average value ratio, usually equal to 2.85.

The first three criteria are based on finding and comparing the peaks and evaluating the spectral components against the largest peaks in the spectral diagram. The last three criteria are based on statistical values and compare these indicators with predefined thresholds, except for criterion 4, which uses the highest spectral peak amplitude.

The criterion associated with the final decision, called the MFD, is responsible for the fusion of all the algorithms. This is done by checking whether the weighted average of the algorithms' results exceeds a predefined threshold. It is defined as follows:

$$\text{MFD} = \boldsymbol{r}^T w \tag{6.29}$$

where $\boldsymbol{r}^T = [r_1, r_2, r_3, r_4, r_5, r_6]$ represents the results vector for the six criteria and $\boldsymbol{w} = [w_1, w_2, w_3, w_4, w_5, w_6]^T$ is the normed weights vector used to calculate the final result, with the sum of the weights equal to 1.

The weights and values used for the parameters of the criteria presented above were established from real experiments under different conditions. They have indicated that scanning in real conditions is most likely performed in the case of a very low SNR and a non-uniform substance temperature. The values used are $\boldsymbol{w} = [0, 04, 0, 09, 0, 14, 0, 29, 0, 24, 0, 20]^T$. The final decision is defined as follows:

$$\text{Decision} = \begin{cases} 1(\text{detected}), & \text{MFD} \geq \text{T} \\ 0(\text{not detected}), & \text{MFD} < \text{T} \end{cases} \tag{6.30}$$

where T is the decision threshold, usually equal to 0.28.

Therefore, if the weighted average of the results is higher than the threshold, the MFD algorithm decides that the substance is detected; otherwise, the MFD decides that the substance has not been detected, as illustrated in Fig. 6.2.

In addition, two separate analyzes were performed to determine the optimal weights. In [5], the optimization was performed using the genetic algorithm (GA), obtaining the weights $\boldsymbol{w}_{GA} = [0.08, 0, 0, 0.21, 0.15, 0.56]^T$. Also, the clonal selection algorithm (CS) was applied in [6], obtaining the weights w $\boldsymbol{w}_{CS} = [0.02, 0.01, 0.03, 0.28, 0.37, 0.29]^T$. In the following comparative analysis, the algorithms will be called MFD, MFD-GA, respectively MFD-CS.

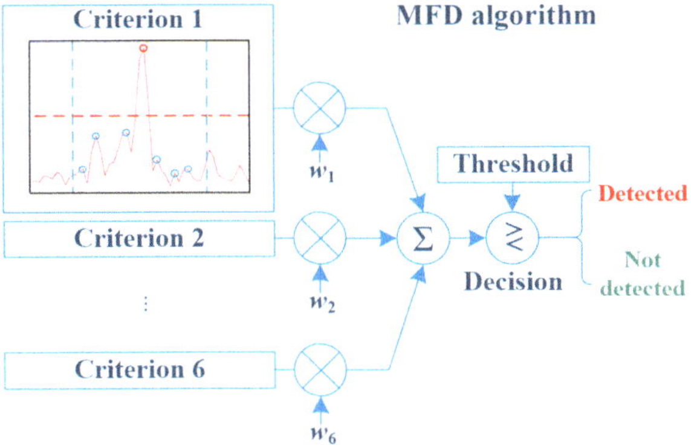

Fig. 6.2 Illustration of the multi-criteria Fourier analysis decision (MFD) algorithm [3] (© 2020 IEEE. Reprinted, with permission, from C. Monea, G.V. Iana, S. Ionita, M. Oproescu, Multi-criteria decision algorithm for NQR signal detection, 2020 International Conference on Applied Electronics (AE))

6.3 Comparative Analysis of the Signal Processing and Signal Analysis Algorithms

The purpose of this section is to compare the signal processing and analysis algorithms in the same conditions (data sets and signal-to-noise ratio). The comparative analysis is performed by simulation, with data sets and noise conditions generated based on the models presented previously and using the dedicated functions in from the simulation environment, respectively. To this end, the analysis procedure followed is initially presented and the performance indicators used to evaluate the selected solutions are established. Then, the implementation of the simulation program in the software development environment is described, and finally the results of the comparative analysis are presented.

6.3.1 Analysis Procedure and Performance Indicators

Taking into account operational aspects and the factors that influence the NQR detection, the following comparative analysis indicators are suggested: processing time and classification performance. The estimation of the processing time is motivated by operational reasons.

Table 6.1 details the classification performance. TP is the number of true positive results, FP is the number of false positive results, and TN and FN represent the number of true and false negative results, respectively. The indicators in the table are determined based on the confusion matrix defined in Table 2.1.

This comparative analysis will make its own benchmark regarding the performance indicators to which new solutions developed for NQR detection can relate. In this regard, the procedure aims at generating the ROC curve in different conditions. The ROC curve plots of the probability of detection (P_d) vs. that of false alarms (P_f or 1-specificity); it can be used to find the optimal detection threshold. For each optimal threshold, the values of the classification performance indicators will be presented. Several simulation scenarios are defined to evaluate the solution's robustness to:

- noise—the algorithms will be evaluated at the SNRs -25 to -50 dB;
- substance temperature variation—the same SNRs will be used, but with the substance's temperature lower by 10 K.

Figure 6.3 illustrates the procedure followed.

In the first stage, the data sets are generated and stored. This stage starts by creating the ET, according to Eq. (5.11) and the parameters in Tables. 5.1 and 5.2. The TNT response was chosen, due to the availability of the data necessary to estimate the model used by the statistical algorithms. One data set is composed of 1000 ETs, the first half containing the useful signal (indicating the substance's presence), and the other only noise (indicating the substance's absence). Each ET is composed of 10 echoes of 2 56 samples. The sampling frequency is 5 MHz. The

Table 6.1 Performance indicators [7–9]

Crt. no.	Indicator	Description	Interpretation	Definition
1	TPR	True Positive Rate/recall/sensitivity/detection probability	The proportion of true positive results correctly identified from all positive results	TP/(TP + FN)
2	TNR	True Negative Rate/selectivity/specificity	The proportion of true negative results correctly identified from all negative results	TN/(TN + FP)
3	PPV	Positive Predictive Value/Precision	The proportion of true positive results correctly identified	TP/(TP + FP)
4	NPV	Negative Predictive Value	The proportion of true negative results correctly identified	TN/(TN + FN)
5	FNR	False Negative Rate/miss rate	The proportion of false negative results misidentified from all positive results	FN / (FN + TP)
6	FPR	False Positive Rate/fall-out	The proportion of false positive results misidentified from all negative results	FP/(FP + TN)
7	FDR	False Discovery Rate	The predicted proportion of false positive results	FP/(FP + TP)
8	FOR	False Omission Rate	The predicted proportion of false negative results	FN/(FN + TN)
9	ACC	Accuracy	Classification accuracy	(TP + TN) /(TP + TN + FP + FN)

(continued)

Table 6.1 (continued)

Crt. no.	Indicator	Description	Interpretation	Definition
10	F_1	F_1 score	Harmonic average of accuracy and sensitivity	2TP/(2TP + FP + FN)
11	J	Informedness	Youden index	TPR + TNR—1
12	AUC	Area Under Curve	Degree of separation between the detection classes	$\int_0^1 \mathrm{TPR}\left(\mathrm{FPR}^{-1}(x)\right)dx$

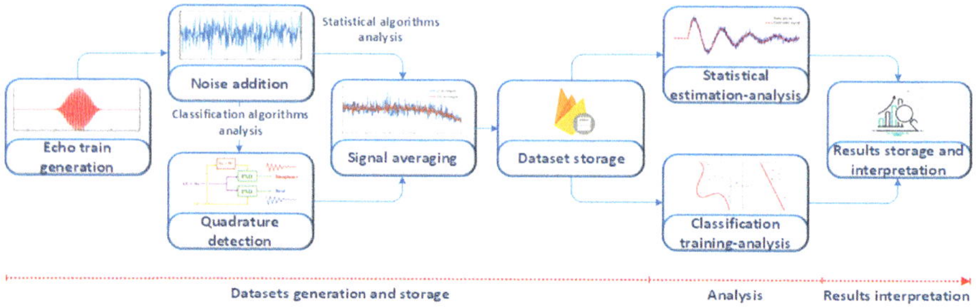

Fig. 6.3 The comparative analysis procedure [10]. (Reprinted from J. of the Franklin Inst., vol. 357, issue 17, C. Monea, A review of NQR signal processing and analysis techniques, Pages 13,085–13,124, Copyright 2020, with permission from Elsevier [OR APPLICABLE SOCIETY COPYRIGHT OWNER])

simulation is performed with one spectral line, with the frequency 848.0907 kHz. The pre-whitening noise is created using the coefficients from [11]. The substance temperatures are 287 K and 297 K, respectively.

After generating the echo train, the noise is added. A complex additive white Gaussian noise is used, generated so as to achieve the desired SNR, defined as follows:

$$\mathrm{SNR} = 10\log_{10}\frac{\sigma_s^2}{\sigma_w^2}[\mathrm{dB}] \tag{6.31}$$

where σ_s^2 and σ_w^2 are the useful signal and noise variances, respectively. Given that their average values are zero, then the variances are, in fact, the powers of the two signals.

The statistical algorithms process the entire ET, while the classification algorithms (including MFD algorithms) operate on the demodulated echo. Thus, in the latter case, after the addition of noise follows the quadrature detection on the frequency 848.1 kHz. This detection allows the visualization of spectral components with

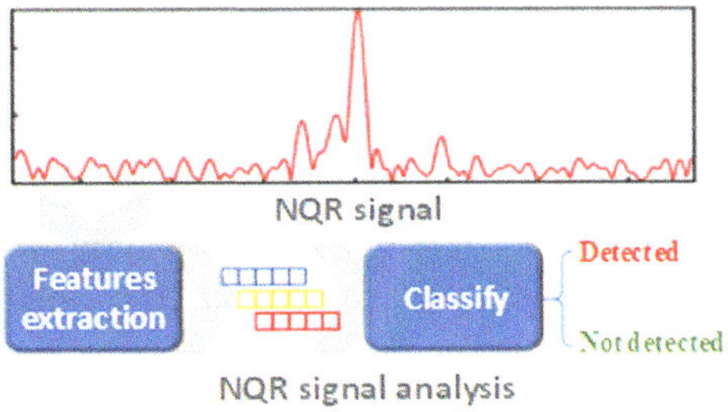

Fig. 6.4 Analysis of the NQR response signal for classification

different frequencies from the resonance frequency, but also the deviation of the excitation frequency from the resonant frequency associated with the temperature of the substance.

Each data set is generated by averaging 4 ETs or all the echoes in the case of the classification solutions. The signal-to-noise ratio defined after adding the noise is used for obtaining a common reference for the statistical and classification solutions.

During the second stage, in the case of statistical algorithms, the analysis starts by estimating the unknown parameters array $\overline{\boldsymbol{\theta}} = \left[T\,\boldsymbol{\beta}^T\,\eta^T\right]$. In the case of classification algorithms, with the exception of MFD algorithms, the analysis starts by training the algorithms using supervised learning. This process extracts the features from the simulated signal's spectrum in order to perform the classification, as illustrated in Fig. 6.4. The confusion matrix is computed for every decision threshold. The ROC curve is calculated in 15 points.

Lastly, the results are stored and interpreted, including the arrays estimated by the statistical algorithms, to allow further analysis.

6.3.2 *Implementation of Comparative Analysis*

The analysis program was implemented in the integrated development environment Spyder 3.3.6 (IDE), in Python 3.7.3. The IDE runs in the docker environment Anaconda 4.7.10, where the following software libraries have been installed to implement the program:

- NumPy 1.16.4—support (mathematical functions and operators) for working with large or multi-dimensional arrays and matrices;
- Matplotlib 3.1.0—for displaying plots;
- Tensorboard 1.14.0—for displaying plots;
- Jsonpickle 1.2—for storing/reading data in/from JSON text files (serialization/deserialization);
- SciPy 1.2.1—for scientific calculations, including optimization, linear algebra, integration, interpolation and signal processing;
- Scikit-learn 0.21.2—for the implementation of ML algorithms used for signal classification (except NN);

- Keras 2.2.4—for the implementation of the NN algorithm;
- Pythonnet 2.4.0—for calling the MFD algorithm developed in C#, in the simulation program written in Python.

The program has a linear structure that follows the steps described in Fig. 6.3.

The **data set generation function** first checks if the data file exists, and if so, reads them from it together with other simulation parameters; otherwise, the function generates data from an initial array representing the echo train. It is created using a function that returns the *m*th echo generated based on Eq. (4.10), to which the Kaiser window with the parameter $\beta = 100$ was applied. The time t is discretized, with the sampling frequency $f_s = 5$ MHz, and the factor $\overline{C}$ is created based on Eq. (6.2) and the coefficients described in [11]. The echo train generating function takes the echo provided by the function described above and positions it in the data array according to its index, at multiples of 256 samples.

The **function for generating and adding noise** takes as input parameters the current echo train and the desired SNR expressed in dB. Firstly, it calculates the noise power based on the input signal power and the desired SNR, as follows:

$$\sigma_w^2 = \frac{\sigma_s^2}{10^{SNR/10}} \tag{6.32}$$

where σ_s^2 is determined using the *var()* function, provided by the NumPy library. Then, the function generates a complex noise vector with normal distribution, the length of the input signal vector and the power σ_w^2, and adds it to the current echo train.

In the case of signals that do not contain the NQR response, it was necessary to simulate the echo train using a random signal with the power equal to that of the useful signal. Otherwise, the noise generation function would have related to a null array, in which case the variance would have been equal to the SNR value, according to the implementation, and the resulting sequence would have had much greater amplitude than in the presence of the useful signal, as illustrated in Fig. 6.5.

The plots show the initial signals, without noise, with dash lines. As seen, the amplitude of the noise sequence is much higher in the absence of the echo train. To obtain similar sequences, the echo train is replaced with a random signal with the power equal to that of the useful signal, approximately $43 \cdot 10^{-4}$ AU.

Quadrature detection (or down-conversion) is applied only in the analysis of the classification algorithms and MFD, because they process the demodulated echo. This is performed on the frequency $f_e = 848.1$ kHz, approximately equal to the excitation frequency, and the resulting complex spectrum will be centered on the reference frequency. From the point of view of signal processing, quadrature detection involves the steps illustrated in Appendix 1. The first stage is the generation (synthesis) of the carrier, representing a complex sinusoid with the (intermediate) frequency f_e. The sequence is defined as follows:

$$f_i[n] = e^{-2\pi i \frac{f_e}{f_S} n} \tag{6.33}$$

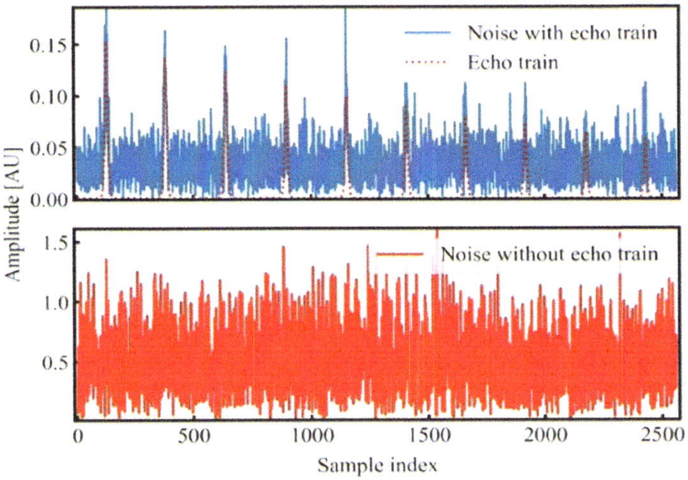

Fig. 6.5 Addition of noise to sequences with (top) and without (bottom) the echo train (SNR = − 5 dB) (AU—arbitrary units)

where f_e is the demodulation frequency, and f_s is the sampling frequency. Then, the echo signal is mixed, resulting spectral components at the frequencies $f_{echo} + f_e$ and $f_{echo} - f_e$, where f_{echo} represents the NQR response frequency (of an echo). Only the last term is of interest, because it is the baseband representation of the complex signal, and it is extracted by low-pass filtering. A decimator filter with finite pulse response, length 64 and cut-off frequency $f_t = 100$ kHz was used. The decimation factor (F_d) is calculated as follows:

$$F_d = \frac{f_s}{2f_t} \tag{6.34}$$

and the sampling frequency after decimation ($f_{s,dec}$) is defined as:

$$f_{s,dec} = \frac{f_s}{F_d} \tag{6.35}$$

Thus, the decimation is performed with a factor of 25, resulting in a sampling frequency of 200 kHz. The size of the echoes that make up the echo train is multiplied by F_d to obtain echoes of 256 samples after decimation.

Averaging is performed every 4 arrays (ETs) for the statistical algorithms, by summing them and dividing by 4. In the case of the analysis of the classification algorithms and MFD, it is done by summing all the echoes within an echo train and performing the division every 4 trains.

The **storage of the generated data** is done in JSON (JavaScript Object Notation) formatted text files, generated with the Jsonpickle software library. This allows data to be stored and formatted so that no further processing is required when reading the files to bring the data into the format recognized by the Python language or the libraries used.

The **data analysis** consists in evaluating the classification performance and the processing time for the investigated algorithms. The first objective involves generating the 15-point ROC plot, which means evaluating a data set 15 times to apply the algorithms studied with variable decision thresholds, recording the results obtained and generating the confusion matrix for each decision threshold. It will be used in the data interpretation stage to determine the performance indicators described in Table 6.1. The ROC analysis function is illustrated in Fig. 6.6.

Figure 6.6 applies to one data set specific to a simulation scenario, and this process is repeated for all data sets and scenarios.

The second objective involves measuring the execution time for each algorithm studied. It measures both the estimation and training time (where applicable). The analysis time for a new data set is also measured. This was implemented using the *timeit* function, based on the Python *timer* class, which allows the evaluation of the execution time of a program sequence with the possibility to repeat the measurements a predefined number of times to reduce the influence of the central processing unit's load on the measurement. In each case 100 measurements were performed, the result representing the average value obtained. Also, when calling the function, it was necessary to specify the *globals()* parameter in order for the evaluated program sequence to use the global variables of the simulation program.

The statistical estimation involves the evaluation of all echo trains in a data set to estimate the vector of unknown parameters $\overline{\boldsymbol{\theta}} = \left[T\boldsymbol{\beta}^T\boldsymbol{\eta}^T\right]$ associated with each of them. In the case of RETAML, the estimation also implies calculating the vector $\hat{\boldsymbol{\alpha}}$. The estimation consists in going through all the values from the search domains associated with the three unknowns, calculating the cost function in each case, storing the results, and, finally, choosing the (optimal) values that maximize/minimize it. This is a search in a huge three-dimensional space, the number of values to be checked being equal to the product of the number of values in the three search domains, which in turn is multiplied by the number of spectral lines estimated. To reduce

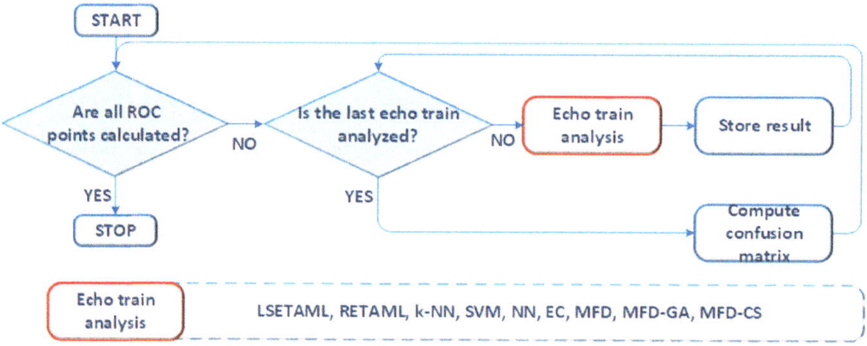

Fig. 6.6 ROC analysis function flowchart

the estimation time, it is proposed to perform three one-dimensional searches. The following search order is used: T-β-η, and each new search uses the previously identified optimal parameters which maintained throughout the value space [11, 12], as illustrated for LSETAML, with signal-to-noise ratio of −5 dB and one spectral line, in Fig. 6.7. This greatly reduces the estimation time. The following search domains are used:$\beta = [0.005, 0.1]$, $T = [292, 302]$ K and $\eta = [2, 4] \cdot 10^{-4}$, which are evaluated in 5 estimation (search) steps; this value is considered sufficient for the evaluation of the algorithms. In practice, the domains are determined based on previously available information, such as the temperature range and decay constants specific to the scanned substance. In the case of RETAML, the assumed amplitude vector $\overline{\boldsymbol{\kappa}}$ is chosen 20% smaller than the real one. Also, the secant method was implemented as a numerical approximation solution for calculating $\hat{\boldsymbol{\alpha}}$ [13].

The analysis itself is based on the data estimated in the previous step to calculate the matrix $\boldsymbol{H}_{\overline{\theta}}$, by means of which the statistical tests described in Eqs. (6.11) and (6.13) are performed. As defined, they return complex results, greater than 1 and have higher as the estimation error is smaller, which makes it difficult to perform the ROC analysis. To avoid this situation, the statistical tests have been modified to provide real results that are lower than 1, as follows:

$$\mathrm{T}\left(z_{NM}, \hat{\boldsymbol{\alpha}}\right) = 1 - \frac{1}{\left|\frac{\|z_{NM}\|_2^2}{\left\|z_{NM} - \boldsymbol{H}_{\bar{\theta}}\hat{\boldsymbol{\alpha}}\right\|_2^2}\right|} \tag{6.36}$$

for RETAML, and

$$\mathrm{T}(z_{NM}) = 1 - \frac{1}{\left|(2NM - 1)\frac{z_{NM}^{*}\boldsymbol{H}_{\bar{\theta}}\boldsymbol{H}_{\bar{\theta}}^{\dagger}z_{NM}}{z_{NM}^{*}\left(\boldsymbol{I} - \boldsymbol{H}_{\bar{\theta}}\boldsymbol{H}_{\bar{\theta}}^{\dagger}\right)z_{NM}}\right|} \tag{6.37}$$

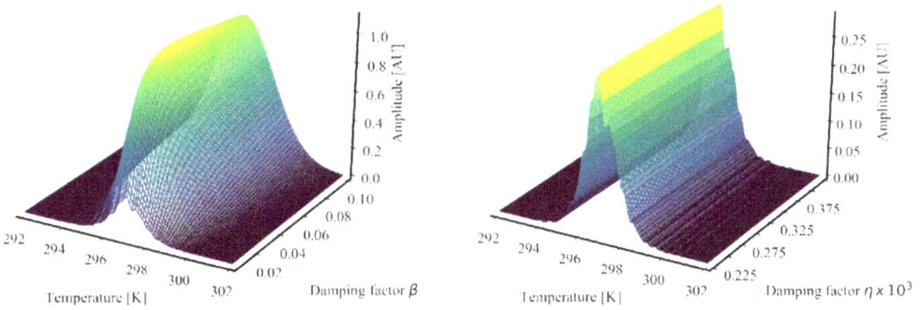

Fig. 6.7 Value of the LSETAML cost function vs. T and β (left) and T and η (right), respectively (AU—arbitrary units) [10]. (Reprinted from J. of the Franklin Inst., vol. 357, issue 17, C. Monea, A review of NQR signal processing and analysis techniques, Pages 13,085–13,124, Copyright 2020, with permission from Elsevier [OR APPLICABLE SOCIETY COPYRIGHT OWNER])

for LSETAML.

The analysis of the classification solutions includes the classifiers training. The training is required for algorithms to learn to distinguish the signals that contain the NQR response from those that contain only noise. Training is done through supervised learning and involves the following sub-stages:

- Data pre-processing

In order to implement the training, the input data is first processed by evaluating all the echo trains within a data set in order to extract and store the signal features and label the sequence, respectively; the results are saved in separate arrays. The following labels were used: "0"—response signal absent and "1"—response signal present.

The features presented in [1, 2] have been proposed for the analysis of ammonium nitrate, but as this study is focused on the simulated TNT signal. Therefore, changes have been made to the features described above. These are presented in Table 6.2, which describes the signal features as implemented.

TNT has a variation coefficient of −0.151 kHz/K for the 848.1 kHz spectral line [14]. An 1024-point FFT was used to calculate the spectral diagram, thus reaching the spectral resolution of 97.65625 Hz, necessary to calculate the features presented in Table 6.2. Also, a rectangular window was used for the FFT calculation, because the use of another type of window would broaden the spectral lobe of the NQR

Table 6.2 Signal features used by the classification algorithms

Crt. no.	Name	Description
1	Feature 1	Amplitude of the spectral component at the reference frequency[a] (848.1 kHz)
2	Feature 2	The sum of the amplitudes of the spectral components in the band 847.1–849.1 kHz
3	Feature 3	The sum of the amplitudes of the spectral components in the band 844.6–851.6 kHz
4	Feature 4	Frequency of the highest spectral peak in the band 823.1–873.1 kHz
5	Feature 5	Frequency of the spectral peak closest to the reference frequency
6	Feature 6	Spectral amplitude at frequency C_5
7	Feature 7	Ratio between the spectral component's amplitude that is located 1.5 kHz left of C_5, and C_6
8	Feature 8	Ratio between the spectral component's amplitude that is located 1.5 kHz right of C_5, and C_6
9	Feature 9	C_1/S_{ref}
10	Feature 10	C_2/S_{ref}
11	Feature 11	C_3/S_{ref}
12	S_{ref}	Spectral average in the bands 833.6 ± 10.5 kHz and 862.35 ± 10.75 kHz

[a]The quadrature detection frequency (848.1 kHz) was used as a reference and not the excitation/response frequency (848.0907 kHz). The values of the two frequencies are very close.

response, although it would reduce the spectral losses (spectral leakage—parasitic spectral components). It was observed that the spectral losses do not influence the detection and the rectangular window was chosen to obtain a narrower spectral lobe, closer to that obtained in real experiments.

- Data splitting

The feature and label arrays are divided into the training and the test sets, respectively. The first set is used for the actual training, and the test set is utilized for checking the algorithms' classification performance using data that have not been used in learning. For this, the *train_test_split()* function of the Scikit-learn library was used, with the percentages of the training–testing sets of 50% and activating the *stratify* parameter to obtain a balanced training set containing the sequences "0" and "1" in the proportions in which they exist in the original data set (in this case, 50%). The data sets are also obtained by randomly choosing feature sequences, improving the accuracy of algorithms by reducing bias (the inability of the classifier to classify correctly due to the limitations of the model used) [15].

- Data scaling

The last stage of data preparation is the normalization of the features, because it is imposed by the k-NN and NN classifiers. The normalization was implemented using the MinMax technique [16], available in the Scikit-learn library. The normalization outputs features sequences with zero average and unit variance. Although not necessary, standardized data are used to test all classifiers, for the simplicity of the implementation.

- Actual training

The *fit()* functions corresponding to each classifier were used. In the case of k-NN, the principal component analysis was used to extract the most significant 5 features, which were used for training and analysis. The classifiers training is done individually. The characteristics of the classifiers used are the following:

- k-NN—classifier with 7 neighbors (7-NN);
- SVM—polynomial kernel;
- NN—11 neurons in the input layer, 5 neurons in the hidden layer and 1 neuron in the output layer, 100 training epochs, dropout value 0.3, 20% validation data set percentage, input and hidden layer activation function—ReLU (Rectified Linear Unit), output layer activation function—sigmoid, Adam optimization algorithm;
- EC—collection of 20 decision trees.

Also called prediction or recognition, the data analysis with classification algorithms is implemented using the *predict_proba()* function in the Scikit-learn library, corresponding to the investigated classifiers. The function is called only once for analysis of the entire test data set and returns an array with the probabilities associated with the two detection classes for each element in this set. This vector is used by the *confusion_matrix()* function in the same library to generate the confusion matrix. When generating the matrix, an additional verification is made to see if the returned probabilities exceed the decision threshold. The ROC curve is created following this procedure, by varying the decision threshold, within the function described in Fig. 6.6. In the case of the k-NN algorithm, the ROC curve is not obtained by varying the number of neighbors, as indicated in [17]; instead, the resulting detection probabilities are compared with a variable decision threshold, as in the case of other classifiers. Thus, the resolution of the ROC curve is no longer limited to the maximum number of neighbors (in this case, 7). The MFD algorithm is developed in the C# language, and the analysis program is written in Python. In order to be able to use the MFD algorithm within it, it was necessary to use the Pythonnet library. This allows the inclusion of the MFD program as an object and to simply call it in the analysis program. The analysis is performed on the test set for the classification solutions, while for the other algorithms it is performed on the entire data set.

The **results storage** is performed in the form of JSON text files, generated with the Jsonpickle library. The estimated data, the classifiers models obtained after training and the confusion matrices related to each ROC point are saved, in order to be able to repeat the analysis without performing again the estimation and training, respectively, and to allow the subsequent interpretation of the results.

The **interpretation of the results** is done in a separate program. It reads the results saved after the simulation and calculates the classification performance indicators; then, it stores them in separate matrices. In order to determine the area under the ROC curve, the Scikit-learn function *auc()*was used, which receives as input parameters the arrays related to the x and y axes (FPR, respectively TPR). The optimal detection threshold was calculated based on the Euclidean distance between the current ROC point and that of a perfect detection (0, 1), with unitary TPR and zero FPR.

6.3.3 Comparative Analysis Results

In order to perform the simulation, 12 data sets were generated. An Alienware Aurora R4 computer with the following technical specifications was used:

- Operating system: Ubuntu 19.04;
- Central processing unit: Intel Core i7-3960X, operating frequency 3,3 GHz, 6 cores, 12 threads;
- Graphical processing unit: Nvidia GeForce GTX 690, 3072 CUDA cores, 4 GB GDDR5 memory, 512-bit memory bus;
- RAM: 32 GB;
- Storage: SSD, capacity 1 TB.

The estimation and analysis for the statistical algorithms took around 3 h/data set. The training and analysis time for classification algorithms was max. of 1 min/data set. The ROC estimation and analysis of the statistical algorithms was performed on the Alienware system, and that of the classification algorithms (including MFD) was performed on a computer with Windows operating system, because MFD is implemented in the C# language and it was not possible to call it in simulation program in Linux/Ubuntu. The following figures show the ROC plots for the statistical algorithms in different noise conditions.

The results are similar, showing a rapidly decreasing performance with the increasing noise level. RETAML is superior to LSETAML. Below −35 dB, the algorithms cannot be applied in practice due to their poor performance. At lower SNRs, the ability to classify the signal becomes comparable to random guessing. The following figures show the ROC plots for the classification algorithms in different noise conditions (Figs. 6.8, 6.9, 6.10, 6.11, 6.12 and 6.13).

These solutions have increased robustness to noise and are superior to the statistical techniques. Furthermore, they can be applied when the signal-to-noise ratio is about 15 dB lower compared to the statistical algorithms (compare Fig. 6.12 with Fig. 6.8).

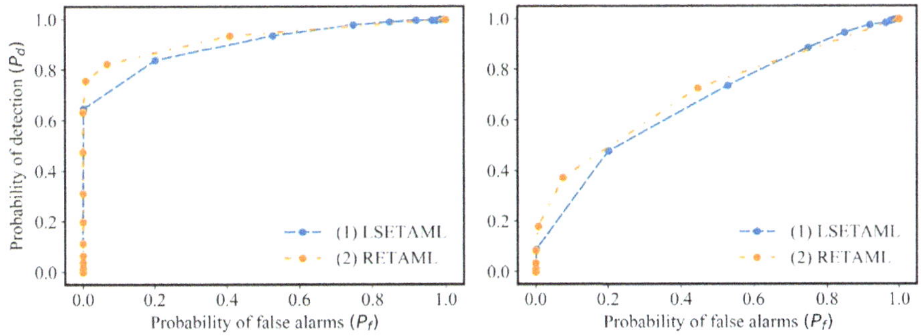

Fig. 6.8 Statistical algorithms ROC curves at the SNRs −30 dB (left) and −35 dB (right) [10]. (Reprinted from J. of the Franklin Inst., vol. 357, issue 17, C. Monea, A review of NQR signal processing and analysis techniques, Pages 13,085–13,124, Copyright 2020, with permission from Elsevier [OR APPLICABLE SOCIETY COPYRIGHT OWNER])

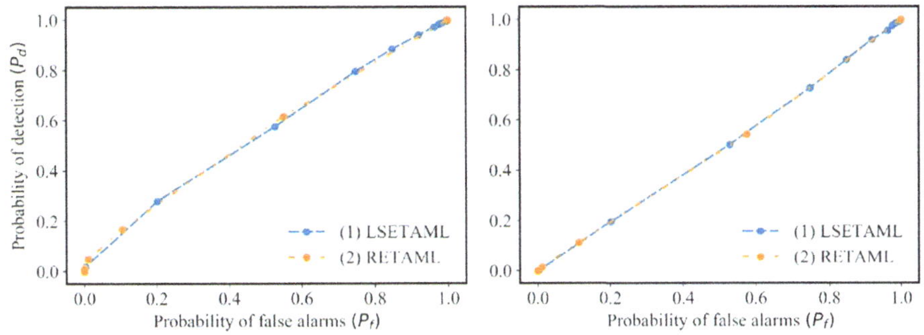

Fig. 6.9 Statistical algorithms ROC curves at the SNRs −40 dB (left) and −50 dB (right) [10]

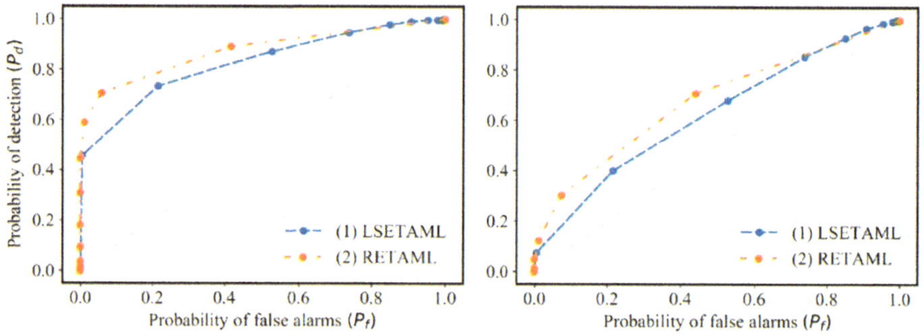

Fig. 6.10 Statistical algorithms ROC curves with a temperature offset of 10 K, at the SNRs −30 dB (left) and −35 dB (right), respectively [10]

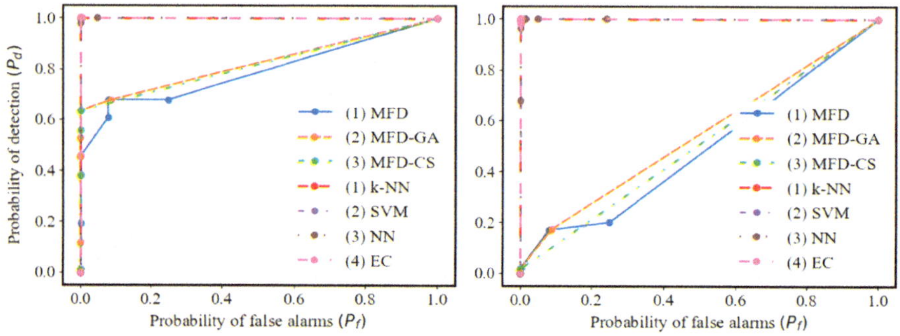

Fig. 6.11 Classification algorithms ROC curves at the SNRs −30 dB (left) and −35 dB (right)

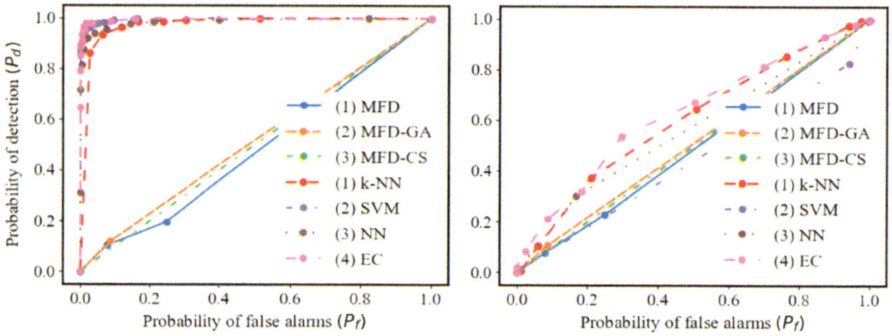

Fig. 6.12 Classification algorithms ROC curves at the SNRs of −40 dB (left) and −50 dB (right)

The following tables present the values of the performance indicators (in percentages) for the studied algorithms (Tables 6.3 and 6.4).

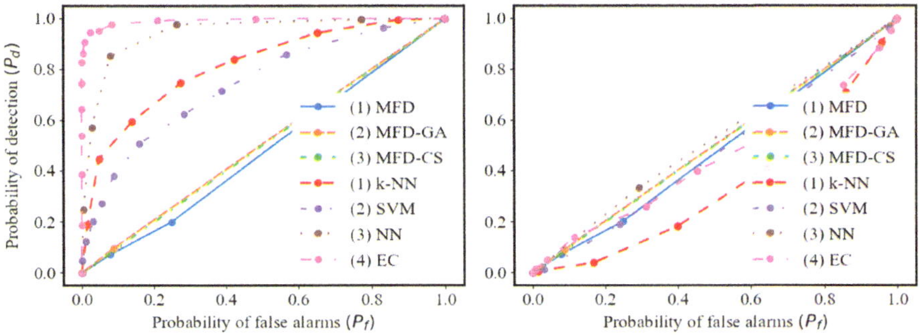

Fig. 6.13 Classification algorithms ROC curves with a temperature offset of 10 K, at the SNRs −45 dB (left) and −50 dB (right)

Table 6.3 Performance indicator values for a SNR of −40 dB (*LS* LSETAML, *RE* RETAML)

Crt. no.	Performance indicator	LS	RE	MFD	MFDGA	MFDCS	k-NN	SVM	NN	EC
1	TPR	57.6	61.8	19.8	1.2	100	93.6	97.6	94.0	98.0
2	TNR	47.2	44.8	75.2	91.2	0	93.6	96.8	96.0	98.4
3	PPV	52.1	52.8	44.3	57.6	50.0	93.6	96.8	95.9	98.3
4	NPV	52.6	53.9	48.3	50.8	0	93.6	97.5	94.1	98.0
5	FNR	42.4	38.2	80.2	88.0	0	6.4	2.4	6.0	2.0
6	FPR	52.8	55.2	24.8	8.8	100	6.4	3.2	4.0	1.6
7	FDR	47.8	47.1	55.6	42.3	50.0	6.4	3.1	4.0	1.6
8	FOR	47.3	46.0	51.6	49.1	0	6.4	2.4	5.8	1.9
9	ACC	**52.4**	**53.3**	**47.5**	**51.6**	**50.0**	**93.6**	**97.2**	**95.0**	**98.2**
10	F_1	54.7	56.9	27.3	19.8	66.6	93.6	97.2	94.9	98.1
11	J	4.8	6.6	−5.0	3.2	0	87.2	94.4	89.9	96.4
12	AUC	54.7	54.9	48.0	51.6	50.0	97.4	99.6	98.6	99.5
13	Optimal threshold	0.79	8e-4	0.07	0.07	0	0.72	0.50	0.43	0.28

It is observed that for the classification algorithms, the optimal detection threshold increases with the SNR's decrease. The classification algorithms have a better robustness to noise and substance temperature offset than the statistical ones. EC achieves the best detection performance.

One of the most important indicators from an operational point of view accuracy (ACC), and its values are highlighted in the previous tables; the aim is to have the highest possible value for ACC.

For clarity, Table 6.5 presents the detection accuracy values (in percentages) for the studied algorithms, calculated at the optimal thresholds. The table reconfirms that EC outperforms other solutions. The lower accuracy of the classification algorithms

Table 6.4 Performance indicator values for a SNR of −40 dB and a temperature offset of 10 K (*LS* LSETAML, *RE* RETAML)

Crt. no	Performance indicator	LS	RE	MFD	MFDGA	MFDCS	k-NN	SVM	NN	EC
1	TPR	56.8	61.6	17.4	12.2	100	95.2	91.2	94.4	97.6
2	TNR	47.4	45.0	75.2	91.2	0	90.4	92.4	91.6	98.8
3	PPV	51.9	52.8	41.2	58.0	50.0	90.8	92.3	91.8	98.7
4	NPV	52.3	53.9	47.6	50.9	0	94.9	91.3	94.2	97.6
5	FNR	43.2	38.4	82.6	87.8	0	4.8	8.8	5.6	2.4
6	FPR	52.6	55.0	24.8	8.8	100	9.6	7.6	8.4	1.2
7	FDR	48.0	47.1	58.7	41.9	50.0	9.1	7.6	8.1	1.2
8	FOR	47.6	46.0	52.3	49.0	0	5.0	8.6	5.7	2.3
9	ACC	**52.1**	**53.3**	**46.3**	**51.7**	**50.0**	**92.8**	**91.8**	**93.0**	**98.2**
10	F_1	54.2	56.8	24.4	20.1	66.6	92.9	91.7	93.0	98.1
11	J	4.1	6.6	−7.0	3.4	0	85.5	83.6	85.9	96.4
12	AUC	54.0	54.2	46.8	51.7	50.0	97.5	97.4	97.9	99.8
13	Optimal threshold	0.79	8e-4	0.07	0.07	0	0.28	0.36	0.36	0.28

Table 6.5 Algorithms detection accuracy (*LS* LSETAML, *T* temperature, *RE* RETAML)

ACC	SNR [dB]									
	T [K]	LS	RE	MFD	MFD GA	MFD CS	k-NN	SVM	NN	EC
−50	297	48.7	48.4	49.0	51.0	50.0	57.0	44.4	56.6	62.0
−45		49.2	49.0	48.3	51.6	50.0	69.8	75.0	73.6	77.4
−40		52.4	53.3	47.5	51.6	50.0	93.6	97.2	95.0	98.2
−35		63.8	64.0	54.6	54.3	50.8	100	100	99.8	100
−30		81.8	87.8	80.1	79.6	81.8	100	100	100	100
−25		99.2	99.5	99.8	100	100	100	100	100	100
−50	287	49.9	49.8	47.7	50.2	50.0	38.4	47.6	52.0	47.4
−45		50.5	50.6	47.6	50.5	50.0	73.6	67.0	88.6	96.0
−40		52.1	53.3	46.3	51.7	50.0	92.8	91.8	93.0	96.4
−35		57.7	63.5	54.4	54.2	50.9	100	100	99.8	100
−30		76.0	82.4	79.9	79.4	82.3	100	100	100	100
−25		96.7	97.6	99.9	100	100	100	100	100	100

at lower SNRs is explained by the fact that the echo train's (with noise) variance is close to that of the signal that contains only noise, making it more difficult to discriminate the two situations.

Figure 6.14 illustrates the learning curves during the training of the NN.

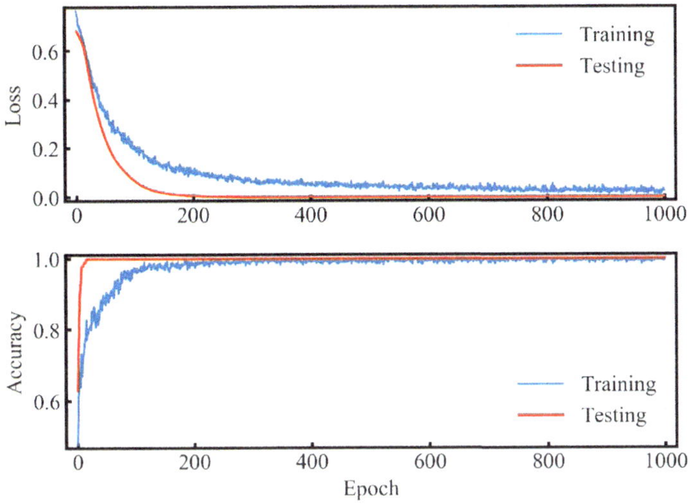

Fig. 6.14 Cost function and accuracy during the training of the NN [10]. (Reprinted from J. of the Franklin Inst., vol. 357, issue 17, C. Monea, A review of NQR signal processing and analysis techniques, Pages 13,085–13,124, Copyright 2020, with permission from Elsevier [OR APPLICABLE SOCIETY COPYRIGHT OWNER])

Table 6.6 Algorithms processing times

Crt. no	Algorithm	Estimation/training time (ms)	Analysis time (ms)	Final processing time (ms)
1	LSETAML	4916.32	0.51	4916.83
2	RETAML	3027.97	0.21	3028.18
3	MFD/-GA/-CS	0	0.90	0.90
4	k-NN	10.28	1.34	1.34
5	SVM	7.14	0.30	0.30
6	NN	1263.62	1.84	1.84
7	EC	23.97	3.86	3.86

The evolution of the cost (loss) function is good, since it is desired to have an abrupt exponential decrease that approaches 0. The bottom plot shows the evolution of accuracy during training and it is observed that after about 400 epochs the two converge to 1.

The processing time shows if the algorithms can be applied in practice (e.g., scanning at a security terminal cannot take too long). The processing time was measured on a computer with Windows operating system, because the MFD algorithm developed in C# could not be called under Linux in the simulation program; also, it was desired to make the measurements in the same conditions.

Table 6.6 shows the results for each algorithm. It is observed that the statistical algorithms have an estimation time hundreds of times longer than the training time of the classification solutions (except NN). It is mentioned that the MFD algorithm does not require estimation/training and is able to analyze the data directly. The total processing time is much longer for the statistical techniques than for classification solutions. This is due to the need for the algorithms to estimate model parameters for each new echo train, an operation that is time consuming. Thus, the classification solutions are more adequate in an operational environment.

The results of the comparative study showed the better performance of the classification algorithms compared to the statistical ones regarding the detection of the NQR signal. Therefore, as indicated in the previous chapter, future research will focus on classification algorithms specific to the field of machine learning in order to achieve an acceptable detection performance in real scenarios.

6.4 Conclusions

The chapter aimed to perform a comparative study of the SPA algorithms applied in NQR applications, in order to verify and create its own benchmark for performance indicators. Given that multi-pulse sequences are the most commonly used, this work was focused on echo train-type response signals.

The study focused on several significant solutions used over time, consisting in statistical algorithms and classification solutions specific to machine learning, respectively. The statistical algorithms were chosen because they represent a large part of the existing solutions and are used for developing other techniques. LSETAML and RETAML were proposed for the study. Also, the classification algorithms were selected because the field of machine learning is very active. The following solutions were proposed: SVM, k-NN, EC and NN. In addition, in order to give a higher degree of novelty to the comparative analysis, a new multi-criteria algorithm (MFD) was proposed and presented in three optimized variants. A brief presentation of the 9 algorithms was made and their mathematical aspects were also described.

The analysis procedure and performance indicators were defined for the comparative study. The analysis procedure aims to perform the simulation steps to measure performance indicators. Since the purpose of the algorithms is to discriminate the presence of the NQR signal from noise, the indicators that show the classification performance have been chosen. Also, the estimation of the processing time is motivated by operational reasons. Numerous simulation program implementation aspects were presented, such as how to generate the data sets, noise, to perform quadrature detection, signal averaging, data analysis and results interpretation. The description was made following the defined analysis procedure, providing details on the libraries used, the implementation mode (including code sequences), as well as the signals obtained after each simulation stage.

The comparative analysis of the 9 algorithms involved the processing of 12,000 echo trains, grouped in 12 data sets associated to different temperature and noise scenarios. The data generated at 6 noise levels in the range –25 to −50 dB and with a temperature offset of 10 K were processed in order to evaluate the algorithms' robustness to different noise levels and the resonance frequency deviation due temperature variation of the substance. The results were presented in tabular form and ROC curves, representing the values of the performance indicators at the optimal detection threshold and the processing times, respectively. The classification solutions outperform the statistical ones and MFD in all simulated conditions. Also, the statistical solutions take the longest time for data processing. Among the machine learning algorithms, the best detection performance is obtained with EC, and the shortest processing time is given by SVM. Therefore, the investigation of machine learning algorithms is proposed to improve detection in real conditions. This study can be used as a reference for evaluating the performance of newly developed solutions.

Appendix

Quadrature Detection

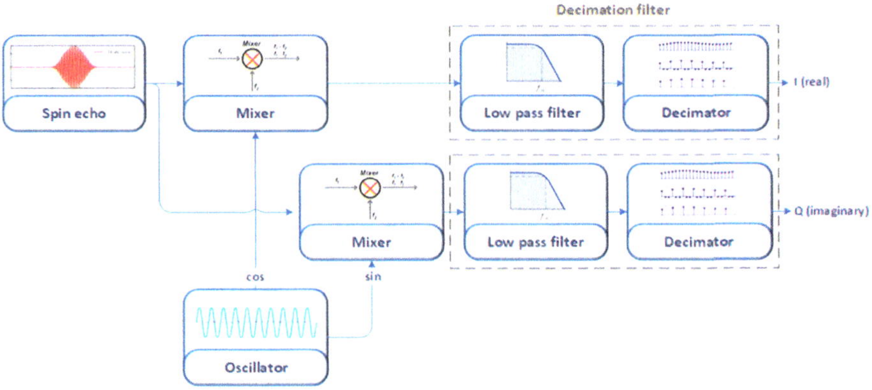

References

1. Cardona, L.R.: Nuclear Quadrupole Resonance System for Landmine Detection in Antioquia. Ph.D. thesis, National University of Colombia (2017)
2. Cardona, L., Itozaki, H., Jiménez, J., Vanegas, N., Sato-Akaba, H.: Spectral descriptors and supervised classifier for ammonium nitrate detection in landmines by nuclear quadrupole resonance. J. Magn. Reson. **305**, 104–111 (2019)
3. Monea, C., Iana, G.V., Ionita, S., Oproescu, M.: Multi-criteria decision algorithm for NQR signal detection. In: 25th International Conference on Application Electronics (2020)
4. Boslaugh, S.: Statistics in a Nutshell, 2nd edn. O'Reilly Media (2012)
5. Oproescu, M., Iana, G.V., Monea, C.: Application of genetic algorithm for optimization of NQR signal detection. In: 11th International Conference on Electronics Computers and Artificial Intelligence, 1–4 (2019)

6. Iana, G.V., Monea, C., Oproescu, M.: The use of the clonal selection algorithm for NQR signal detection optimization. In: 11th International Conference on Electronics Computers and Artificial Intelligence, 1–4 (2019)
7. Fawcett, T.: An introduction to ROC analysis. Pattern Recogn. Lett. **27**(8), 861–874 (2006)
8. Powers, D.M.W.: Evaluation: From Precision, Recall and F-factor to ROC, Informedness, Markedness & Correlation, Technical report. Flinders University (2007)
9. Wikipedia.: Receiver operating characteristic. Internet: https://en.wikipedia.org/wiki/Receiver_operating_characteristic. 18 Aug. 2020
10. Monea, C.: A review of NQR signal processing and analysis techniques. J. Franklin Inst. **357**(17), 13085–13124 (2020)
11. Jakobsson, A., Mossberg, M., Rowe, M.D., Smith, J.A.S.: Exploiting temperature dependency in the detection of NQR signals. IEEE Trans. Signal Process. **54**(5), 1610–1616 (2006)
12. Somasundaram, S.D., Jakobsson, A., Gudmundson, E.: Exploiting spin echo decay in the detection of nuclear quadrupole resonance signals. IEEE Trans. Geosci. Remote Sensing **45**(4), 925–933 (2007)
13. Somasundaram, S.D., Jakobsson, A., Gudmundson, E.: Robust nuclear quadrupole resonance signal detection allowing for amplitude uncertainties. IEEE Trans. Signal Process. **56**(3), 887–894 (2008)
14. Mandal, S., Song, Y.Q.: Method and system for applying NQR pulse sequences. Patent, WO2013134474 (2013)
15. Kuhn, M., Johnson, K.: Feature Engineering and Selection: A Practical Approach for Predictive Models. Chapman and Hall/CRC (2019)
16. Scikit-learn, MinMaxScaler. Internet: https://scikit-learn.org/stable/modules/generated/sklearn.preprocessing.MinMaxScaler.html. 18 Aug. 2020
17. Majnik, M., Bosnić, Z.: ROC analysis of classifiers in machine learning: a survey. Intell. Data Anal. **17**(3), 531–558 (2013)

Chapter 7
Analysis of Nuclear Quadrupole Resonance Response Signals

7.1 Description of the Analyzed Signals

The identification of the NQR response signal is a binary classification problem (detected/not detected) and this study aims to propose new signal analysis solutions in order to improve the detection performance. In order to obtain the most relevant results, the algorithms' development is based on real data, which were acquired using a custom-designed spectrometer [1] designed for pulse cNQR detection. The equipment allows excitation using the FID, SLSE and WURST-QCMPG sequences. The latter was used for the general advantages offered by the multi-pulse sequences and the particular advantage of compensating the variation of the substance's resonance frequency with the temperature [2]. A 7-step frequency sweep scan was performed in the band f_e ±2 kHz, where f_e represents the excitation frequency, thus allowing the excitation of the substance when its temperature differs by up to a few degrees from the reference. The data was acquired for the following substances:

- 130 g paracetamol ($C_8H_9NO_2$)—precursor found in drug combinations;
- 150 g sodium nitrite ($NaNO_2$)—toxic substance in large quantities.

The substances have the resonant frequencies of 4.642 MHz (sodium nitrite) and 2.563 MHz (paracetamol) at the temperature of 297 K, and the relaxation times T_1 are 100 ms and 19 s, respectively. Taking into account that they have the frequency-temperature coefficients of −1.6 kHz/K [3] and 0.06 kHz/K [4], then the scan covers a range of approx. ±1 and ±16 K, respectively. The data set used for the development of the signal analysis algorithms consists of 20,000 acquisitions, half of which are obtained in the presence of the substances of interest (positive scans—PS) and half in its absence (negative scans—NS). The acquisitions are also distributed evenly, with 10,000 acquisitions for each substance. In order to obtain the most relevant results, the acquisitions were made in various scanning conditions (scenarios), as follows:

C. Monea and N. Bizon, *Signal Processing and Analysis Techniques for Nuclear Quadrupole Resonance Spectroscopy*, Signals and Communication Technology,
https://doi.org/10.1007/978-3-030-87861-0_7

- At different distances between substances and the excitation/detection coil (placed directly on the coil, at 1, 2, 3 and 5 cm);
- In the presence of sand;
- With the temperature of the substance modified;
- With a detuned resonant detection circuit;
- Using a small substance quantity;
- In the substance's absence.

In the first case, the different distances between the coil and the substance will generate acquisitions with different SNR, reflecting the real situations when the substances inside the scanned objects (luggage, packages or envelopes) can be located at different distances from the coil. In the second case, one obtains acquisitions that contain false spectral peaks due to the noise generated by the inverse piezoelectric effect that occurs in substances with crystalline structures without a center of symmetry [5], such as sand (400 g were used). The data acquired in the third case allows the development of algorithms to discriminate the situations in which the substance's response has a frequency offset due to the variation of its temperature from the reference one. In this scenario, the substance's temperature was lowered by several degrees by temporarily storing in, before the measurements, in a laboratory with a lower temperature than the one in which the spectrometer is located. The situation is also relevant from a practical point of view, reflecting cases where the substance located in the luggage, e.g., has a different temperature than the reference. The fourth case allows the analysis of response signals in the presence of false spectral peaks that may occur when the resonant circuit is slightly out of tune due to metal objects found inside the scanned object. The fifth case allows the analysis of signals in very low SNR conditions. The last scenario allows the discrimination of noise from the substance's response. It is mentioned that part of this data set was used to optimize the MFD algorithm presented in a previous chapter.

The excitation and acquisition system integrated in the spectrometer allows the reception of echo signals after averaging and quadrature detection. A number of 1000 mediations was used, and the detection resulted in 65 complex samples. The spectral bandwidth used for reception is 185 kHz for sodium nitrite and 50 kHz for paracetamol. The time domain signal is determined based on the following equation:

$$y_{mag}(i) = \sqrt{re(i)^2 + im(i)^2} \tag{7.1}$$

where i is the sample index, $re(i)$ is the real value (in phase) and $im(i)$ is the imaginary value (in quadrature). The time domain (magnitude) signal represents the envelope of the averaged echo. The following figures show the NQR response when the substance is placed on the detection coil and in its absence, respectively (Figs. 7.1 and 7.2).

One can observe the spectral peak of the response in the presence of the substance and the noise in its absence, in both time and frequency domains. The spectral signal is calculated in 1024 points using the FFT and mirroring the halves of the spectral diagram to observe the peak of the NQR response in the center and to discriminate

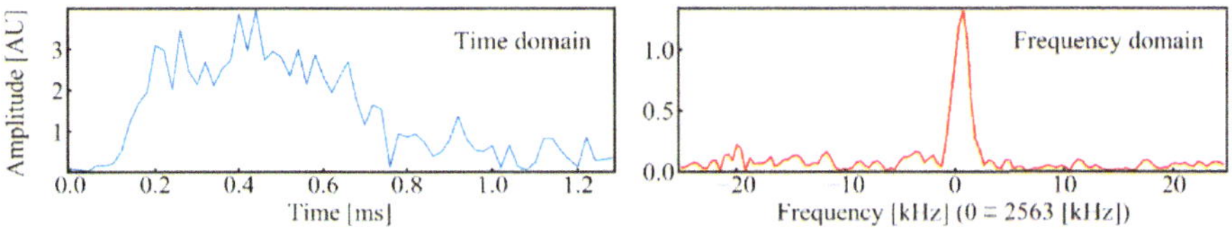

Fig. 7.1 NQR response of paracetamol placed on the detection coil, in time and frequency domains (AU—arbitrary units)

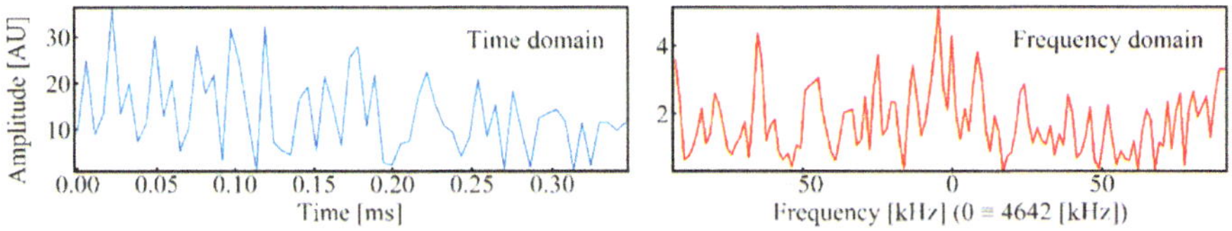

Fig. 7.2 The signal obtained in the absence of sodium nitrite, in the time and frequency domains (AU—arbitrary units)

the slight deviation from the reference (excitation) frequency. Several representative acquisitions made in the above-mentioned scenarios can be found in Appendix 1.

7.2 Defining the Signal Features

The signal analysis algorithms developed in this study aim to discriminate the NQR response from unwanted signals represented by noise and radio frequency interference. In order to achieve this, the use machine learning techniques is desired, which perform the signals classification by associating them to a detection class based on signal features. In order to establish these features, the characteristics of the response signal must be taken into account, and this is a very important stage of applied machine learning [6]. The relevance of the features influences the detection performance achieved by the developed algorithms.

The feature represents a measurable individual property or of an observed phenomenon [7]. This is common to all records (acquisitions) in the data set and is used by the signal analysis algorithm to make the prediction (classification). The property can represent practically anything, from numerical values (e.g., measurements or statistics), arrays or even labels (character arrays) or graphs. Generally, the NQR response can be considered a non-stationary (temporal) time series, characterized by the exponential echo decay, the presence of several response frequencies, short duration and reduced SNR [8]. Given that the discrimination of the substance response's presence is performed by humans primarily by evaluating the spectral diagram, then this is proposed as a feature. The time domain signal is not proposed as a feature for performing the classification, due to the high variability caused by the influence of unwanted signals and the dependence on the excitation and substance sequence, which is less observable in the spectral domain. The form of the NQR response depends both on the number of acquisitions made, through the averaging technique, and on the type of the scanned substance, which has particular relaxation times. These values may differ from one substance to another and, together with the influence of unwanted signals, lead to different signal forms, which requires the collection of a larger data set to achieve good detection performance.

In order to increase the probability of detection and reduce false alarms, some additional features resulting from the visual analysis of the spectra obtained in each scan scenario are proposed. It has been observed that the frequency of the spectral peak response is located in the plot's center, around the reference (excitation) frequency, and its amplitude is maximum. Thus, it is considered relevant to know the frequency of the maximum spectral peak.

In scenarios with low SNR, due to the long distance from the substance to the coil, the amplitude of the response peak becomes comparable to that of the noise. For this reason, it is considered useful to know this fact for discriminating the PS acquisitions from NS. Also, the specific spectral form of the NQR response is analyzed by performing the correlation with a reference spectrum, presented in a previous chapter, obtaining the Pearson correlation coefficient. The reference spectrum is obtained under optimal conditions and can be used to perform the discrimination, taking into account that the NQR response is similar for all substances excited with a single resonant frequency.

In situations when the substance's temperature is changed from the reference or the resonant circuit is detuned, the spectral peak appears broadened, split or containing false components, but is located in the center of the spectral diagram and is noticeable from the noise level. Therefore, it is considered relevant to know the ratio between the average value of the center frequency band and that of the side bands. The center band is defined as $f_e \pm 10$ kHz, and the value is considered sufficient to contain the response of the substance.

The classification algorithms have been applied in the past to discriminate the NQR response of ammonium nitrate and have yielded promising results [9]. In order to improve the detection performance, the 11 signal features proposed in this case are also taken into account.

All these features are accompanied by a label indicating the presence or absence of the response signal. The labels "Detected" were used for PS and "Not detected" for NS, respectively. Two classes are used, because it is not of interest to discriminate the type of substance based on the response, since this is performed by the NQR method, which identifies it based on its unique resonance frequency.

Table 7.1 summarizes the selected features and also describes their purpose in the training of the developed algorithms. In total, 1040 signal features are used, of which 1024 represent the spectral diagram's samples, 4 are defined in this study, 11 are proposed in other works, and 1 represents the detection class associated with each record in the data set.

Table 7.1 Features used in the development of the signal analysis algorithms

Index	Name	Description	Purpose
0	0	Sample 0 of the spectral diagram	Distinguish the noise response signal in different situations
1	1	Sample 1 of the spectral diagram	
...			
1023	1023	Sample 1023 of the spectral diagram	
1024	Spectral maximum frequency (SMF)	Maximum spectral peak frequency	Learning that the resonance frequency is located in the center band
1025	Spectral average to maximum peak (SAMP)	The ratio between the maximum peak amplitude and the spectral average	Distinguishing NS acquisitions by the fact that the peaks' level is comparable to that of noise
1026	Pearson Correlation Coefficient (PCC)	Pearson correlation coefficient with a reference spectrum	Recognition of the distinctive form of the NQR response
1027	Spectral regions average ratio (SRAR)	The ratio between the average of the central band and that of the side bands	Recognition of the response in scenarios with modified temperature, detuned resonant circuit, piezoelectric noise and low substance quantity
1028	C1	Amplitude of the spectral component at the reference frequency (excitation; f_e)	Indication of the NQR response's presence
1029	C2	The sum of the amplitudes of the spectral components in the band f_e ±1 kHz	Distinguish the response signal in the center band, considering the area around the peak
1030	C3	The sum of the amplitudes of the spectral components in the band f_e ±3.5 kHz	Distinguishing the response signal in the extended central band, considering it a wider area around the peak
1031	C4	Frequency of the highest spectral peak in the band f_e ±25 kHz	Noise and RFI recognition
1032	C5	Frequency of the spectral peak closest to the reference frequency	Identification of the response in case of substance temperature offset
1033	C6	Spectral amplitude at frequency C_5	Used to calculate the features C6 and C7

(continued)

Table 7.1 (continued)

Index	Name	Description	Purpose
1034	C7	Spectral amplitude at 1.5 kHz left of C_5, divided by C_6	Recognition of the response signal's shape
1035	C8	Spectral amplitude at 1.5 kHz right of C_5, divided by C_6	
1036	C9	C_1/S_{ref}	
1037	C10	C_2/S_{ref}	
1038	C11	C_3/S_{ref}	
1039	Class	Label indicating the detection class	Indication of the presence or absence of the response signal

The S_{ref} value used to calculate the features C9-C11 represents the spectral average in the bands $[f_e—25, f_e—3.5]$ kHz and $[f_e + 3.5, f_e + 25]$ kHz. The features C1-C11, including S_{ref}, that were initially developed for the NQR response of ammonium nitrate, have been adapted to be applicable to any substance. Thus, the frequencies and bands used were related to the excitation frequency. The spectral diagrams are calculated in 1024 points and taking into account that the spectral widths used at reception are 50 kHz and 185 kHz, then the resolutions obtained are approx. 48 Hz and 180 Hz, enough to calculate the features described in the table.

7.3 Statistical Analysis of Data

An exploratory analysis of the data set is performed in terms of the previously defined signal features, in order to better understand them and to discover new information about them and the relationships between them. Thus, descriptive statistical techniques will be used. Table 7.2 presents some of the values of the features that make up the data set.

For clarity, the previous table shows only a subset of the features used, in the case of the first and last 5 records of the data set. Their relevance is indicated by the fact that it is possible to discriminate the two detection classes by observing the higher values of the features in the case of PS (detected) compared to those obtained in the case of FS (not detected). The maximum peak's spectral frequency (SMF) is shifted from the reference (frequency 0 in the complex spectral diagram), due to the measurement conditions and the FFT resolution. It is observed that the maximum peak's frequency is located in frequency bands located far from the reference in the case of NS, due to noise and the absence of the response signal. The ratio between the maximum peak's amplitude and the spectral average (SAMP) is higher in the presence of the response, whose level is higher than that of the noise, following the averages

Table 7.2 Data set values (*AU* arbitrary units, *D* detected, *ND* not detected)

Acquisition index	0 (UA)	…	1023 (UA)	SMF (Hz)	SAMP	PCC	SRAR	C1 (UA)	…	Class
0	0.49	…	0.51	−904	14.06	0.59	10.62	12.26	…	D
1	0.11		0.11	−904	12.57	0.59	7.93	11.84		D
2	0.36		0.38	−1085	11.52	0.54	10.23	11.07		D
3	0.22		0.15	−1265	10.91	0.53	9.34	10.37		D
4	0.41		0.48	−1085	10.76	0.53	9.57	10.67		D
…										
19,995	1e-3	…	1e-3	−39e3	2.93	0.28	1.03	2e-3	…	ND
19,996	5e-3		4e-3	54e3	3.01	0.28	1.15	1e-3		ND
19,997	9e-3		9e-3	−16e3	2.43	0.22	0.94	4e-3		ND
19,998	6e-3		6e-3	−25e3	2.87	0.32	0.95	8e-3		ND
19,999	6e-3		7e-3	37e3	2.70	0.22	1.08	8e-3		ND

performed during the signal acquisition. As expected, the Pearson correlation index (PCC) and the spectral average ratio (SRAR) are lower in the case of NS due to the lack of response.

The distributions of the values of SMF, SAMP, PCC and SRAR can be easily observed in Fig. 7.3. The histograms of the C1-C11 features and the distributions of the feature values in the form of Boxplot diagrams are presented in Appendix 2.

The histograms show the number of observations with the same values (frequency) for all features. The plots show, in red lines, the 0 and 100% percentiles, respectively the quartiles Q1, Q2 and Q3. The percentages are the values below which certain percentages of the data set's values are located. Their order is: 0, Q1, Q2, Q3 and

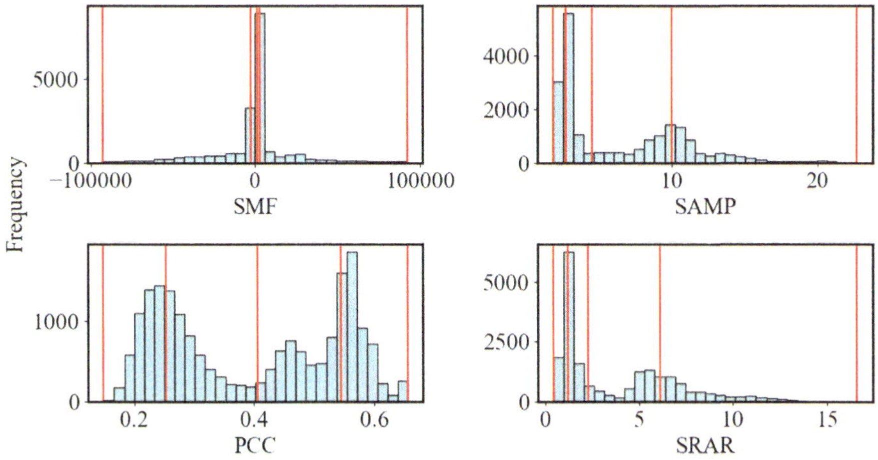

Fig. 7.3 Histograms of the SMF, SAMP, PCC and SRAR features

100%. The symmetrical, unimodal distribution (with only one maximum) of SMF is noticed and is explained by the fact that the frequency of the response is close to the reference, and in the case of NS it is given by the random noise. SAMP and SRAR have bimodal distributions and indicate that most NS give approximately equal values for these features, in the 0–5 and 0–3 ranges, respectively, while PS give higher but more scattered values. The PCC has a multimodal distribution, noting that the NS are grouped at lower values, and the PS at higher values.

The correlations between the features can be observed using the scattering matrix. This is a collection of scatter plots of several variables, illustrated in a matrix form, and is used to find out if the variables are correlated and if the correlation is negative or positive. The points related to the two detection classes are colored separately. The distributions of the values of the investigated features are illustrated on the main diagonal. In Fig. 7.4 the weak correlations between SMF and SAMP, PCC, SRAR are observed, as well as the overlap of the values related to the two classes.

An alternative method of observing the correlations between the features is to use the Heatmap diagram. This is a colored representation of the values of a matrix and has the advantage that it can be interpreted more easily. Figure 7.5 illustrates the correlation matrix for samples 384–640, representing the first 128 samples to the

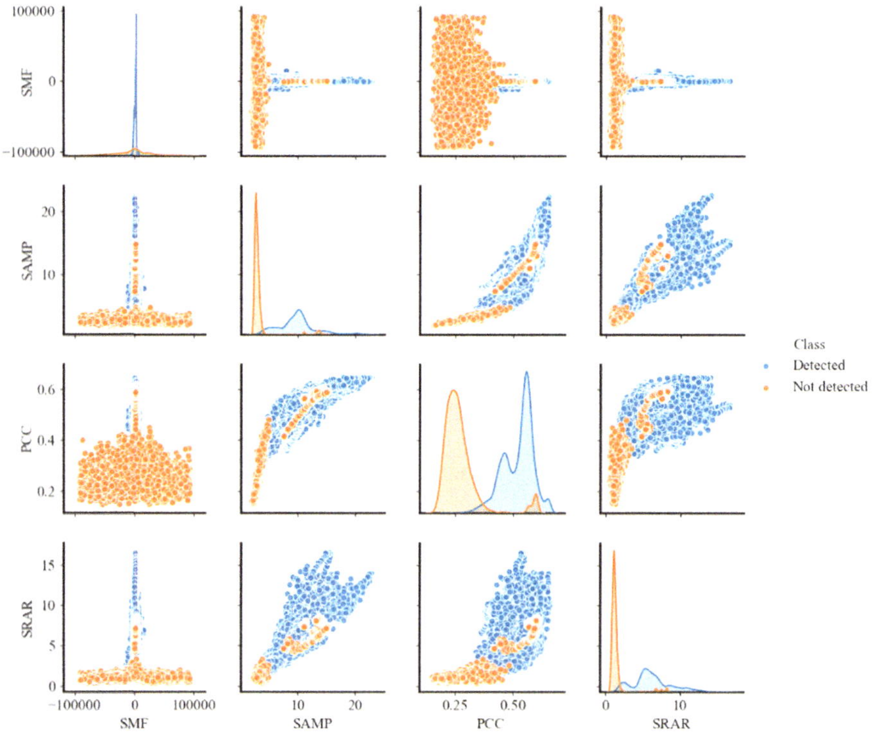

Fig. 7.4 Scatter matrix for the SMF, SAMP, PCC and SRAR features

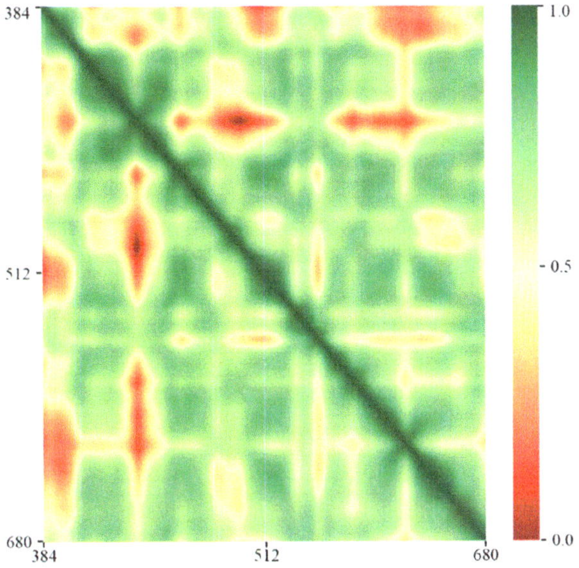

Fig. 7.5 Heatmap diagram of the correlation of samples 384–640 in the spectral diagram

left and right of the reference frequency. The latter is located at sample 512, in the middle of the matrix.

As expected, the samples located in the center band are poorly correlated with the most distant ones, due to the fact that the response signal is located in a narrow band around the reference frequency. There is also a strong correlation between the adjacent samples (along the main diagonal) and a weak one between the side band and central band samples, due to noise. There are also correlations between the central band and certain side bands, indicated by light green areas, which may be due to acquisitions with the substance's response in the form of a split spectral peak, when the substance's temperature is changed from the reference or the resonant circuit is detuned.

7.4 Conclusions

This chapter aimed to present a statistical analysis of the response signals obtained in nuclear quadrupolar resonance spectroscopy. A number of 1040 signal features are proposed for the development of new detection solutions based on machine learning algorithms. A data set composed of 20,000 signal acquisitions in different conditions was created and used generating the feature data set. The descriptive statistical analysis of the data set has brought insight regarding the relationships between the proposed features.

Appendix

Signal Acquisitions in Different Scanning Scenarios

NQR response of paracetamol placed on the detection coil, in time and frequency domains (AU—arbitrary units).

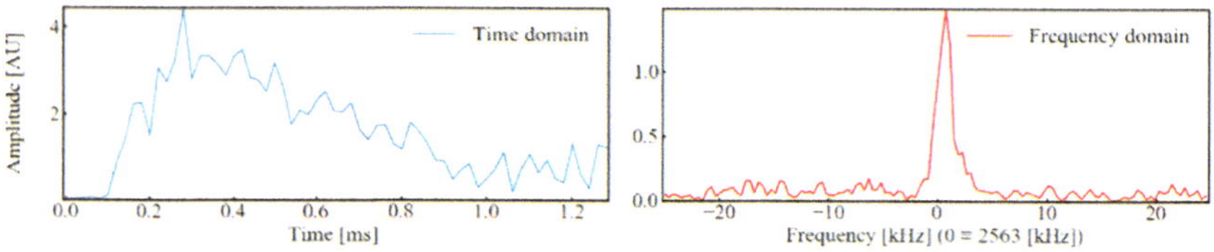

NQR response of paracetamol placed 3 cm from the detection coil, in time and frequency domains (AU—arbitrary units).

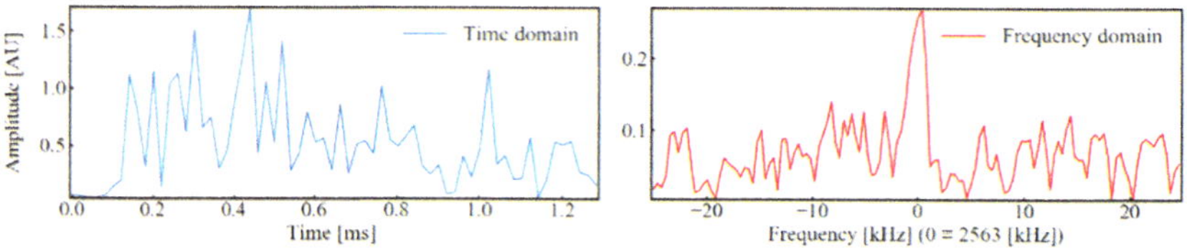

NQR response of sodium nitrite in the presence of sand, in time and frequency domains (AU—arbitrary units).

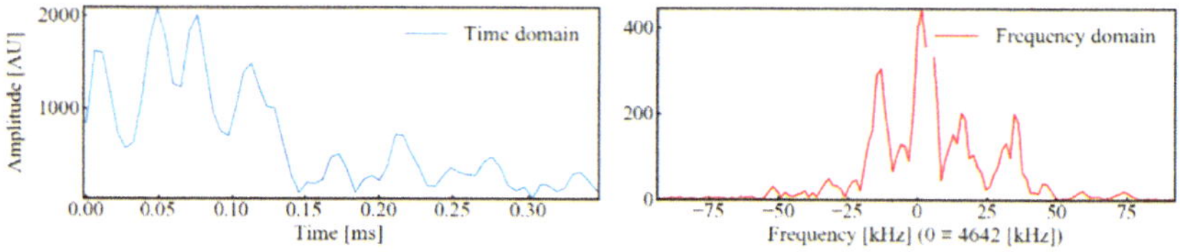

NQR response of sodium nitrite with temperature offset, in time and frequency domains (AU—arbitrary units).

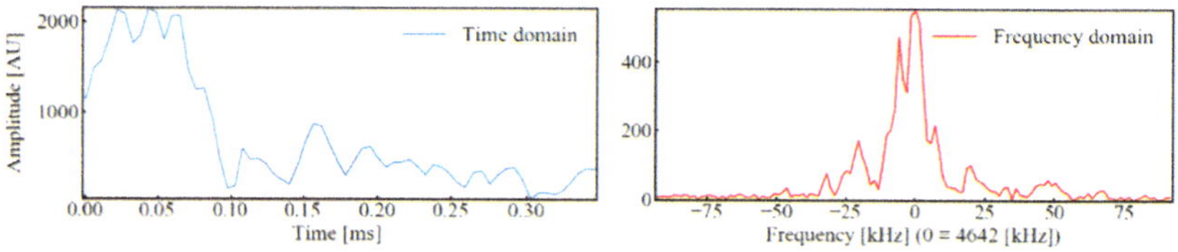

NQR response of sodium nitrite in the case of a slightly detuned resonant circuit, in time and frequency domains (AU—arbitrary units).

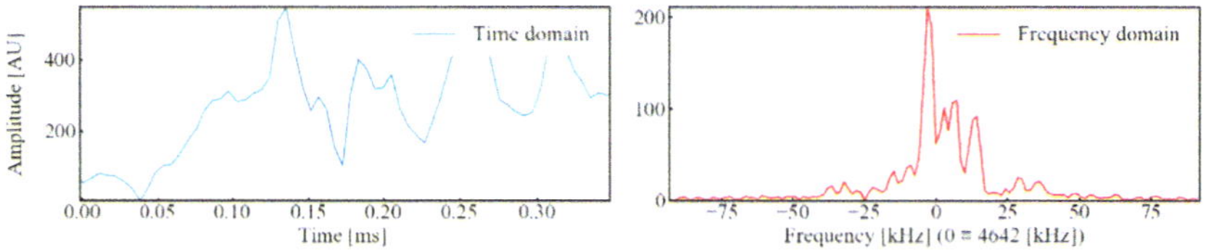

NQR response in the case of one gram of sodium nitrite, in the time and frequency domains (AU - arbitrary units).

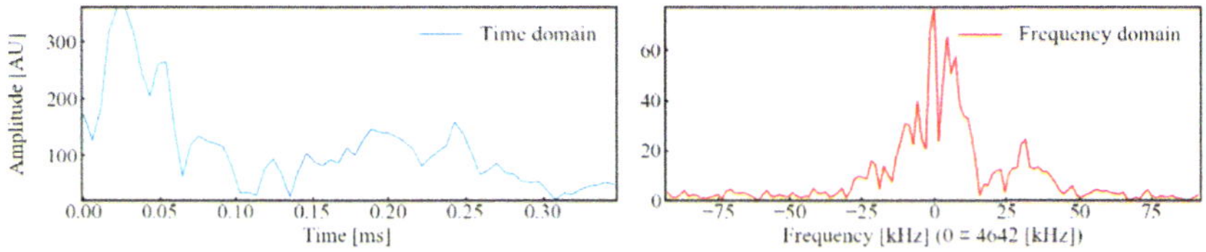

Signal obtained in the absence of sodium nitrite, in the time and frequency domains (AU—arbitrary units).

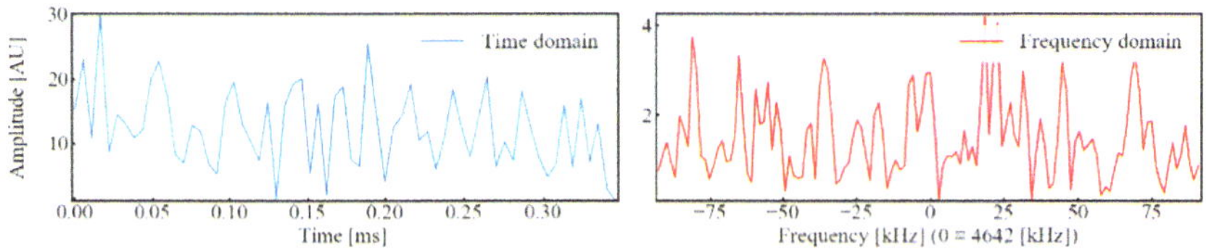

Since the spectrometer uses a planar coil, the SNR will decrease as the distance to the substance decreases [10], as shown in the first and second cases. By placing the substance directly on the coil, the response can be easily distinguished in the form of the demodulated echo's envelope, in the time domain, and as a clear peak in the spectral domain, respectively. The amplitudes obtained are several times smaller than in the first case, because they decrease as the substance is placed further away from the coil. Also, the SNR decreases as the amplitude of the NQR response becomes comparable to the noise level. The sand generates a strong response in the form of false response peaks located in the spectral range. Generally, this type of coherent noise can be reduced using special phase cycling excitation techniques [11, 12], but it is desired to discriminate it using signal analysis algorithms. In the scenario where the substance's temperature is changed from the reference one, it is placed directly on the coil, and the spectral diagram illustrates the splitting of the response peak due to the non-uniform temperature of the substance. If the resonant circuit is detuned, the spectrum has false components in addition to the response of the substance. In the penultimate case, the less clearly defined answer is observed, which is obtained after scanning 1 g of sodium nitrite. The results of the scan performed in the substance's absence show the noise in the time and frequency domains.

Statistical Analysis of the Data Set

Histograms of features C1-C11 and Class.

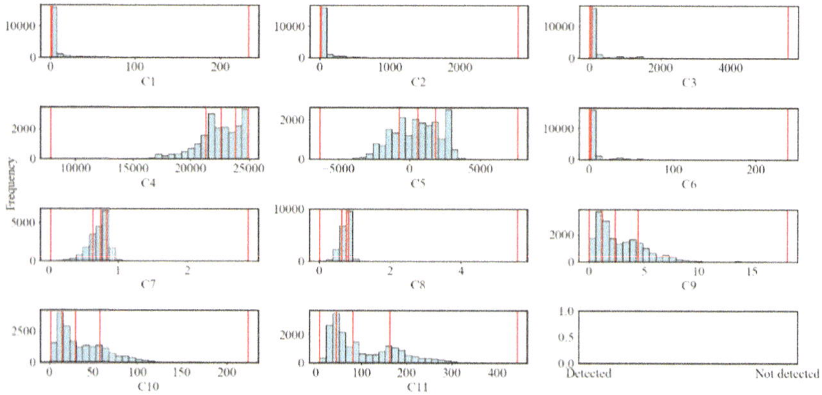

Boxplot diagram of the features SMF, C4 and C5 (left), as well as C1-C3 and C6-C8 (right).

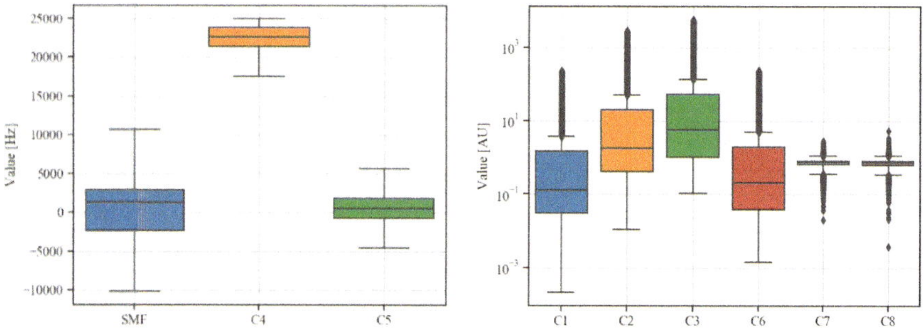

Boxplot diagram of the features SAMP, PCC, SRAR and C9-C11.

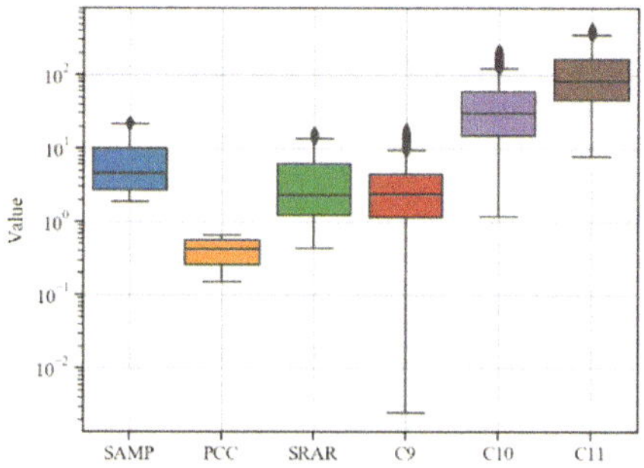

References

1. Monea, C., Iana, G.V., Ionita, S., Ionescu, L.M., Zaharia, S.A., Ilie, S., Bizon, N.: An optimized NQR spectrometer for detection of prohibited substances. Meas. **151**, 107158 (2019)
2. Gregorovič, A., Apih, T.: WURST–QCPMG sequence and 'spin-lock' in 14N nuclear quadrupole resonance. J. Magn. Reson. **233**, 96–102 (2013)
3. Mandal, S., Song, Y.Q.: Method and system for applying NQR pulse sequences. Patent, WO2013134474 (2013)
4. Lužnik, J., Pirnat, J., Jazbinšek, V., Lavrič, Z., Srčič, S., Trontelj, Z.: The influence of pressure in paracetamol tablet compaction on 14N nuclear quadrupole resonance signal. Appl. Magn. Reson. **44**(6), 735–743 (2013)
5. Buess, M.L., Garroway, A.M., Miller, J.B., Yesinowski J.P., Lindquist,, R.P.: Method and apparatus for detecting a target material in a sample by pre-screening the sample for piezoelectric resonance. Patent, US7132942B1 (2005)
6. Ng, A.: Machine Learning and AI via Brain simulations. Internet: https://ai.stanford.edu/~ang/slides/DeepLearning-Mar2013.pptx. 19 Aug. 2020
7. Bishop, C.M.: Pattern Recognition and Machine Learning. Springer (2006)
8. Daniels, D.J.: EM Detection of Concealed Targets. Wiley (2010)
9. Cardona, L., Itozaki, H., Jiménez, J., Vanegas, N., Sato-Akaba, H.: Spectral descriptors and supervised classifier for ammonium nitrate detection in landmines by nuclear quadrupole resonance. J. Magn. Reson. **305**, 104–111 (2019)
10. Ostafin, M., Nogaj, B.: 14N-NQR based device for detection of explosives in landmines. Meas. **40**(1), 43–54 (2007)
11. Rudakov, T.N.: Modification of SSFP technique for the effective detection of NQR signals. Phys. Lett. A **358**(4), 322–326 (2006)
12. Rudakov, T.N., Belyakov, A.V.: Modifications of the steady-state free-precession sequence for the detection of pure nuclear quadrupole resonance. J. Phys. D Appl. Phys. **31**(10), 1251–1256 (1998)

Chapter 8
Development of Signal Analysis Algorithms for NQR Detection

The development starts from existing signal analysis algorithms and aims to modify them for the detection of NQR response signals. Thus, the investigated algorithms are selected, after which they are implemented in the development environment used. Then their comparative evaluation on real data is performed and the best algorithm is chosen. Finally, the proposed algorithm is optimized to obtain the best classification performance and is compared with the results presented in a previous chapter.

8.1 Selection of the Investigated Algorithms

As previously presented, machine learning-specific classification algorithms have been applied in NQR detection and have the potential to achieve acceptable detection performance in real-world scenarios. In order to explain the choice of algorithms, it is necessary to describe their field.

According to Fig. 8.1, machine learning is part of the artificial intelligence (AI) field. The latter is focused on creating systems that can correctly interpret data, learn from it and use the knowledge gained to achieve specific goals and perform tasks through flexible adaptation [1]. Machine learning is the science and art of ensuring the ability of systems to learn from data without being explicitly programmed [2]. In other words, it includes a set of algorithms and training methods to teach them to take decisions based on available data. They have been used in many applications [3–5]. Furthermore, deep learning is a part of machine learning that contains algorithms inspired by the way the human brain processes information and uses many nonlinear information processing layers for the extraction and supervised or unsupervised processing of features, signal recognition and classification [6]. They can automatically generate features for classification, while those specific to machine learning require the manual provision of features, but they generally need longer computing resources and training times.

C. Monea and N. Bizon, *Signal Processing and Analysis Techniques for Nuclear Quadrupole Resonance Spectroscopy*, Signals and Communication Technology,
https://doi.org/10.1007/978-3-030-87861-0_8

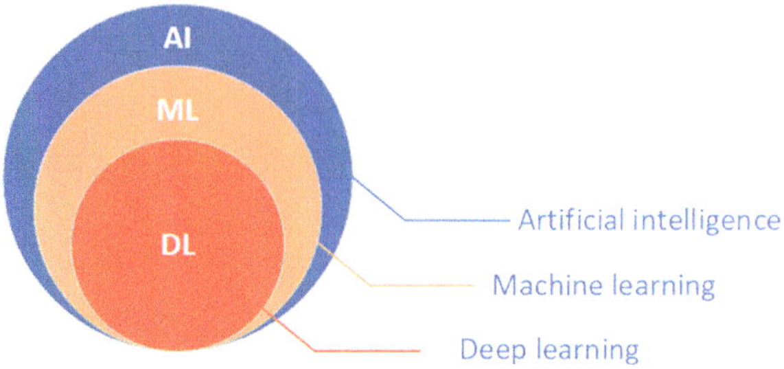

Fig. 8.1 The relationship between artificial intelligence, machine learning and deep learning

The deep learning algorithms have been developed quickly in recent years due to the increase in available data and the scale of algorithms [7]. Their development was also supported by the annual ImageNet Large Scale Visual Recognition Challenge (ILSVRC) competition, which led to the creation of the most advanced image recognition algorithms, such as: LeNet, AlexNet, VGG, GoogLeNet, ResNet [8], SE-ResNet. The latter achieved a classification error of 2.25%, lower than that of a human expert [9]. Generally, these algorithms allow increasing the classification performance when the amount of data increases [7].

The classification algorithms used so far in NQR detection represent a small part of the existing solutions. This study aims to investigate specific deep learning solutions, as they have achieved very good performance in classification and have not been applied in NQR detection yet. In order to obtain the best result, several algorithms used in the classification of one-dimensional (1D) data [10] are selected, which will be evaluated in order to select the best one. The following supervised learning solutions are chosen for study for their high performance and in order to cover several categories of algorithms: DNN, CNN, LeNet-5, AlexNet, VGG-16, ResNet-34, LSTM, CNN LSTM and ConvLSTM. Some of them have been developed for image recognition and will be modified for 1D signals. The following is a brief description of the algorithms studied.

The **deep neural network** is an artificial neural network with several layers of hidden neurons [11]. The output of each neuron is a function of the weighted sum of its inputs (activation function). The DNN identifies the linear or nonlinear relationship that exists between its input and output.

The **convolutional neural network** is a neural network that uses the operation of convolution in at least one of its layers [12] and is inspired by the way the human brain processes images. It is composed of an input and output layer, and several hidden layers that perform the convolution operation with specific filters or kernels, resulting in feature maps, as shown in Fig. 8.2.

The figure illustrates the convolution in the case of an image, and if the signal is one-dimensional, the 1D convolution will be used [13, 14]. The network also includes pooling or subsampling layers, which replace an area of the map with a statistic (maximum value—*Max-Pooling* or average value—*Average-Pooling*), and densely connected layers (fully connected), obtained from a multi-layer perceptron, which requires the flattening of the feature maps. Flattening is the transformation of

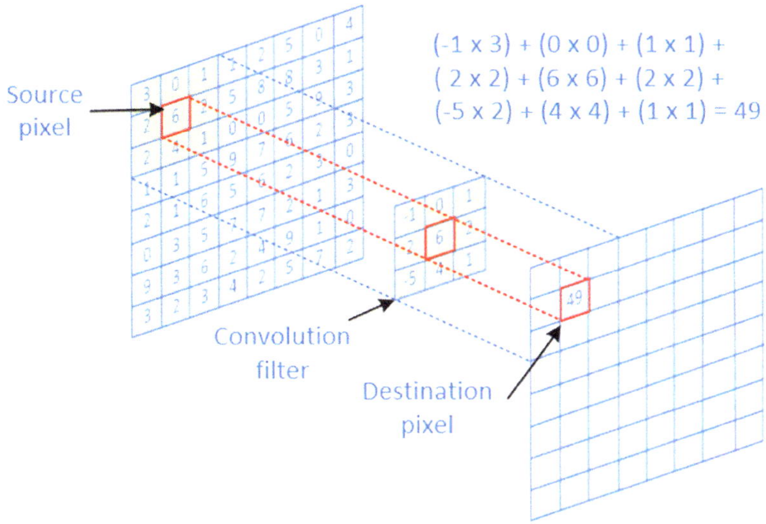

Fig. 8.2 Convolution of an image with a filter

a matrix into an array in order to apply it to a densely connected layer. The structure of a CNN is illustrated in Fig. 8.3.

The figure shows a network that receives at its input an image with three channels (corresponding to the three primary colors—red, green and blue), of size 64 × 64 pixels, and performs the convolution with 24 kernels of size 48 × 48 pixels. Then, the subsampling and flattening operations are performed and the data is passed to a densely connected layer with 256 neurons. Finally, the detection result is provided by the output layer, with 128 neurons. It is mentioned that CNN generates features automatically through the grouping operation, but it can also use manually generated features.

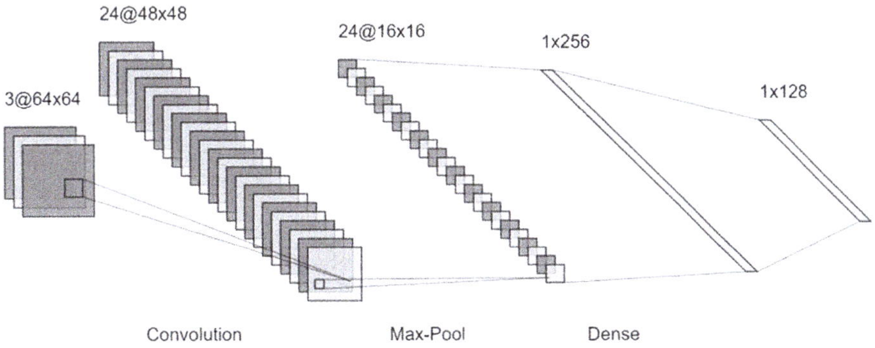

Fig. 8.3 The structure of a convolutional neural network

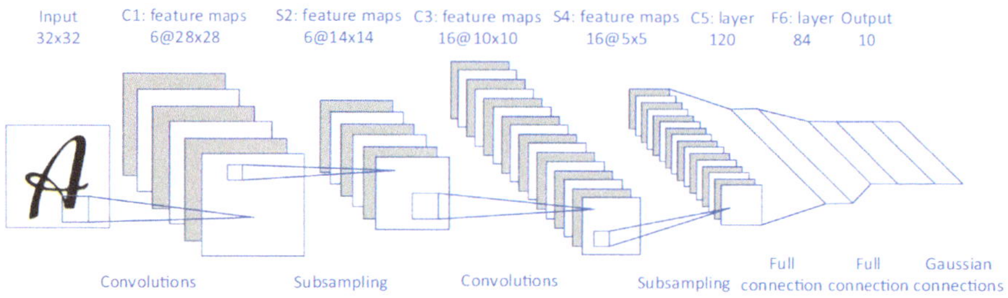

Fig. 8.4 LeNet-5 network architecture

The **LeNet-5 network** was developed for handwriting recognition in low-resolution images [15] and was the first to contain successive layers of convolution and subsampling, followed by fully connected layers. Its architecture is illustrated in Fig. 8.4.

It contains two convolution layers, with 6 filters of 28 × 28 pixels and 14 × 14 pixels, respectively, and two subsampling layers. Finally, 3 fully connected layers with 120, 84 and 10 neurons are used, obtaining a total of 5 layers that give the name of the architecture. The convolutional layers use the hyperbolic tangent activation function.

The AlexNet architecture was the first with 5 convolutional layers, larger inputs and kernels and a semi-linear activation function (ReLU), which won the ILSVRC competition [16]. The architecture is illustrated in Fig. 8.5.

The architecture is divided in two, because at the time of its development the algorithm could not be trained using only a graphics processor, which was limited by the capacity of volatile memory. The AlexNet network is designed for 224 × 224 pixel images and contains 5 convolutional layers, with 11 × 11, 5 × 5 and 3 × 3 pixel filters, respectively 3 fully connected layers. The image also illustrates the convolutional kernel's stride.

VGG-16 (Visual Geometry Group-16) demonstrated the increase of the classification performance with the depth of the network, and is composed of 13 convolutional layers and 3 fully connected layers [17], as illustrated in Fig. 8.6.

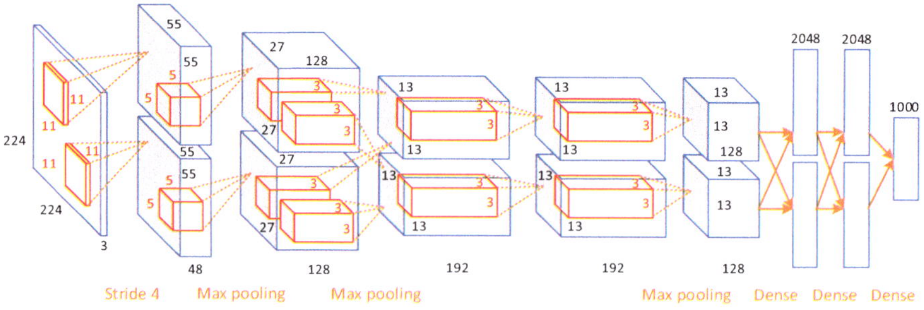

Fig. 8.5 AlexNet network architecture

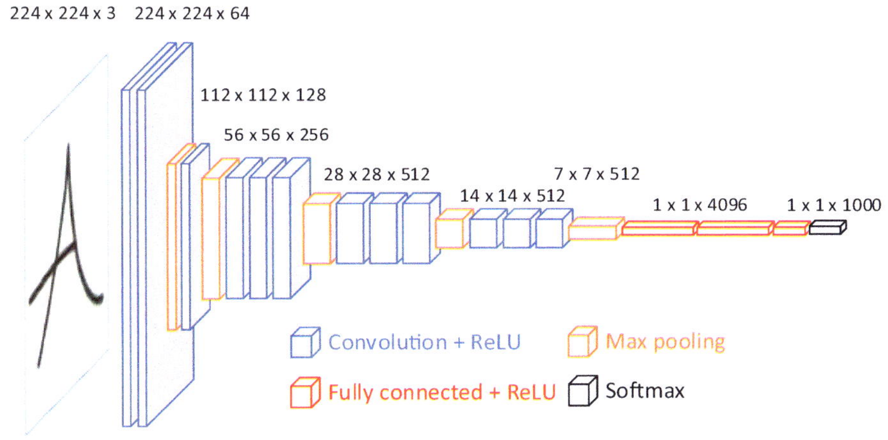

Fig. 8.6 VGG-16 network architecture

The network uses smaller filters than AlexNet, but increases the number of layers to 16. The activation function for the last layer is softmax, which is used in multi-class classification.

Studies of deep architectures have shown that as the depth increases, the accuracy of the classification becomes saturated, after which it decreases rapidly. The **ResNet-34** (Residual Neural Network-34) was the first to solve this problem, allowing the use of deep models (architectures) with shortcut connections [18]. Very deep networks were difficult to train, and the solution to this problem is the introduction of the residual block (which gives the name of the architecture), which involves adding the input to the output of a block composed of several convolutions (shortcut connection). Also, it was possible to decrease the number of filters per layer by using multiple layers.

The **LSTM** (Long Short-Term Memory) model is a type of recurrent neural network that can learn and store long sequences of data. The LSTM architecture is illustrated in Fig. 8.7.

An RNN network processes information over time and composed of several neural networks connected to each other. They are arranged successively along the time axis,

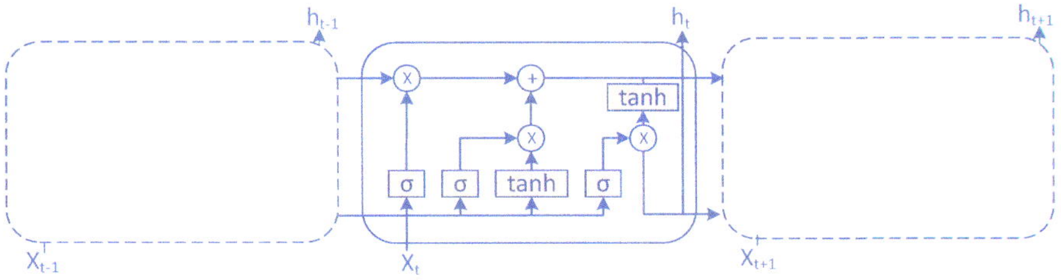

Fig. 8.7 Architecture of a LSTM network

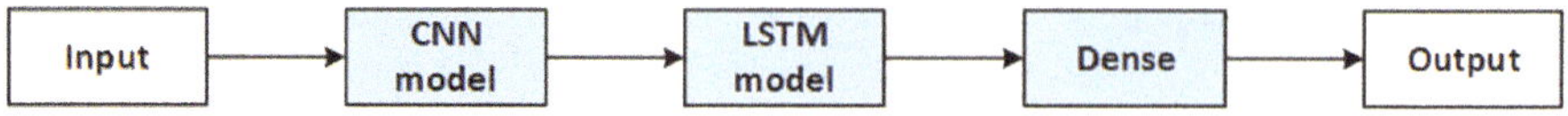

Fig. 8.8 The general architecture of a CNN LSTM network

in the sense that they process the data received at different times. The stored information (internal state) regarding the data prior to the current time is transmitted through the connections between the networks, and the decision taken by the model at the previous time will influence the decision taken at the current time. The LSTM architecture provides a complex structure for each neuron in the hidden layers, which contains gate elements, that command the reading or modification of stored information. These gates are controlled by sigmoid activation functions, which receive both the data at the current time and the stored information.

The **CNN-LSTM** (Convolutional Neural Network Long Short-Term Memory) is a combination of the convolutional neural network and the LSTM netowrk, where the convolutional layers are used to extract the features that are provided to the net recurrent network [19]. Finally, a fully connected layer is used. The general architecture of this model is illustrated in Fig. 8.8.

The **ConvLSTM** (Convolutional Long Short-Term Memory) network differs from the previous one in that it performs convolutions within the LSTM network. Basically, the CNN model is integrated into the LSTM model. Thus, unlike LSTM, which reads data directly to calculate the internal state, and CNN LSTM, which interprets the output of the CNN model, the ConvLSTM network performs the convolution directly in LSTM units [20].

8.2 Implementation of the Signal Analysis Algorithms

The implementation is performed in the integrated development environment Spyder 3.3.6, in the Python 3.7.3 language, running in the docker environment Anaconda 4.7.10. The software libraries used were described in a previous chapter, except that in this case the Tensorflow 2.0.0 library is used to develop machine learning algorithms, which integrates the Keras library. The Pandas 0.25.3 library dedicated to data manipulation and analysis is also used. The system used to run the developed programs is provided free of charge by the Google Colaboratory cloud platform and has the following specifications:

- Operating system: Ubuntu 18.04.3 LTS;
- Central processing unit: Intel Xeon, operating frequency 2,3 GHz, 1 core, 2 threads, cache memory 45 MB;
- Graphical processing unit: Nvidia Tesla K80, 2496 CUDA cores, 12 GB GDDR5;
- RAM: 12.6 GB;
- Storage: capacity 320 GB.

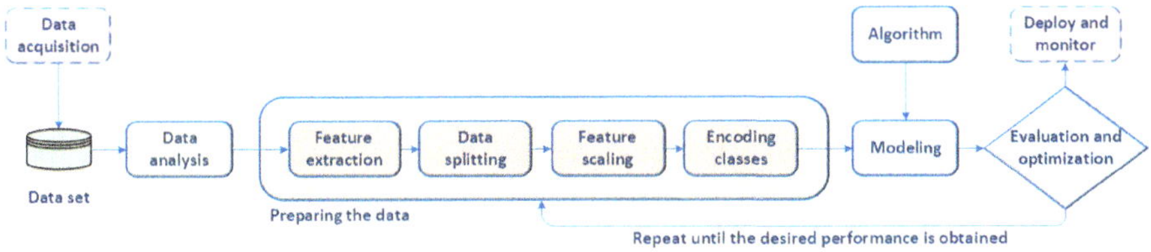

Fig. 8.9 The workflow followed for the development and evaluation of the signal analysis algorithms

The workflow followed for the development and evaluation of the signal analysis algorithms is illustrated in Fig. 8.9.

This workflow is used at every run of the investigated algorithms, except for the blocks marked by dashed lines. The following presents implementation aspects related regarding each step of the workflow illustrated above.

The **data acquisition stage** consists in collecting the records used in the development of the signal analysis algorithms. This was described in a previous section and involved collecting 20,000 acquisitions in several scan scenarios. Following this step, the data set is stored as text files organized in positive (PS) and negative (NS) scans, and further, in scan scenarios. The script file contains hundreds to thousands of acquisitions. In addition, the simulated data sets presented in the previous chapter will be used to evaluate the optimized algorithm.

Reading the data involves going through all the acquisition files, respectively reading and storing the data in a structure that is used in the program. The structure of the text files is given by the excitation and acquisition system integrated in the spectrometer, as follows:

```
#SpinCore Technologies, Inc.
#SF in MHz
#SW in Hz

[HEADER]
@CLASS= RADIOPROCESSOR
@FIRMWARE= 3087
@SPINAPI= 20160816_x86_64
@POINTS= 65
@SF= 4.642000
@SW= 185185.187500

[DATA]
-5787522
-34106918
-5178102
-35028666
4754437
-33648342
...
```

The beginning of a file resulting from a sodium nitrite scan is shown above. The header provides the following information:

- Name of the manufacturer of the excitation and acquisition system;
- Measurement unit for the excitation frequency (SF—Spectrometer Frequency): MHz;
- Measurement unit for the spectral width (SW): Hz;
- System type: radio processor;
- Firmware version: 3087;
- API (Application Program Interface) version: 20160816_x86_64;
- Number of acquired samples: 65;
- Excitation frequency: 4.642 MHz;
- Spectral width: 185,185.1875 Hz;
- Acquired samples expressed in arbitrary units, in pairs, where the first number in a pair represents the real value, and the second, the imaginary value, as they result from the quadrature detection.

Firstly, the number of PS and NS is determined, then all the files are verified and for each file the samples of the time axis (x-axis) are calculated, based on the spectral width value, as well as the amplitude axis (y-axis), by forming complex numbers which are divided by a scaling constant. The latter is determined empirically and is equal to the ratio between the constant 1.499392496135997e8 and the number of averages. For both substances, a number of 1000 averages (scans) are used; thus, the same scaling constant is used. Also, the type of each file is stored as a label indicating the detection class: PS—"Detected", NS—"Not detected".

The data is stored in a Numpy array that contains the following components:

- The number of PS (files);
- The number of NS (files);
- The data array composed of the following data for each file:
  - Time axis samples;
  - Amplitude axis samples;
  - Detection class.

The size of this array is given by the number of files. For efficiency, the data is read only once, and the resulting array is saved in a binary format, and can be read quickly at each program execution. This stage also contains the exploratory statistical analysis of the data presented in a previous chapter.

Feature extraction represents the first preparation (pre-processing) stage of the data and consists in evaluating the array obtained at the previous stage and calculating the 1040 features described in the previous chapter, for each acquisition. The detection class is already included in the array. At the end of this stage, the data obtained are stored in a Pandas structure, because it offers the advantage of easier manipulation and analysis.

Data splitting is necessary to obtain the data sets used to train or test (validate) the algorithms in the evaluation stage, as illustrated in Fig. 8.10. The *train_test_split()* function of the Scikit-learn library was used, with the proportions of the training–testing sets of 80 and 20%, respectively, and activating the *stratify* parameter to obtain balanced sets (containing PS and NS in the proportions in which they exist in the

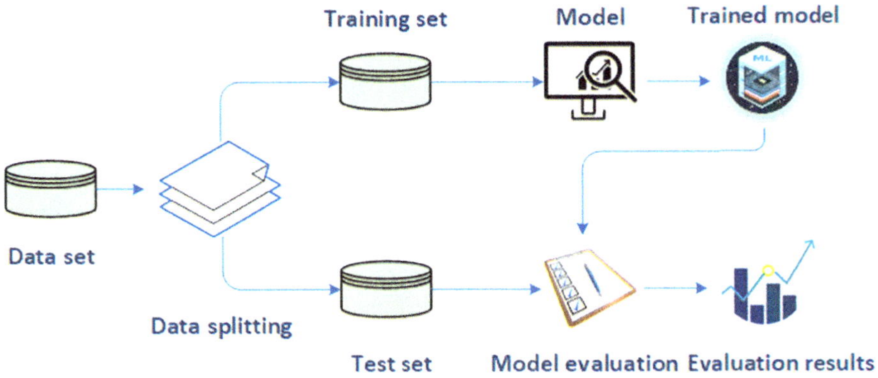

Fig. 8.10 The use of the training and testing sets

initial data set—50%). Thus, the training set consists of 16,000 acquisitions and the test set contains 4000 records. Also, the data sets are randomized using the *shuffle* function from the same library, in order to improve the accuracy of the algorithms by reducing the bias (subjectivity; inability of the classifier to classify correctly due to the limitations of the model used). Also, in order to obtain reproducible results, the same initialization value was used for the random number generator.

The test set is used to evaluate its performance on unknown data, after the algorithm's training is finished.

Normalization is necessary because it generally improves the accuracy of algorithms, the speed of convergence [21] and prevents bias [22]. Generally, this can be done using the min–max method, which scales the data in the given interval, normalizing the averaging value, standardizing (obtaining data with zero average and unit standard deviation), dividing by the maximum value or using a hyperbolic tangent estimator. All features, except for the spectral diagram's samples, are normalized in the range [0, 1] using the min–max technique [23], available in the Scikit-learn library. Thus, if $\boldsymbol{x}$ is the feature array, the normalized array $\boldsymbol{x_n}$ is given by the equation:

$$\boldsymbol{x_n} = \frac{\boldsymbol{x} - x_{\min}}{x_{\max} - x_{\min}} \tag{8.1}$$

where $x_{\min}$ and $x_{\max}$ represent the minimum and maximum value of the data array. The spectral diagram's samples were normalized by dividing by the maximum value, because the min–max method sometimes changes the shape of the diagrams. To prevent finding information from the training set in the test set (data leakage and bias), the normalization is performed separately on the two sets [24, 25].

Modeling involves the use of the set of standardized and encoded features to train the developed algorithm. The training is performed using the *fit* function from the

Tensorflow library, and the Adam [26] training algorithm is used. It aims to stochastically optimize a cost function, by reducing the errors and generalizing the representations learned from the data. There are several cost functions, such as the mean square error (recommended in regression problems), cross-entropy (recommended in classification problems), exponential, Hellinger, Kullback–Leibler or Itakura-Saito. The binary cross-entropy cost function was chosen because it is recommended for this application [27] and does not allow obtaining small gradients when the output values are low. For clarity, it is noted that the difference between regression and classification is that the former represents the prediction of continuous values, while the latter applies to discrete values. The fit function also ensures data shuffling at each training epoch, thus helping to find a local minimum in terms of optimization [2]. The metric used to evaluate the classification performance is accuracy, as it is most indicated in the NQR detection application and is recommended for a balanced data set. As in the data splitting step, the same initialization value for the random number generator used by the Tensorflow library is utilized to obtain reproducible results.

Model evaluation and optimization involves evaluating the performance of the model trained on the test set and optimizing it with the aim of increasing it. The evaluation compares the metrics used to indicate the detection performance (accuracy) with the desired level, then optimizing the algorithm if it does not reach this level. Optimization means changing the model (architecture, hyperparameters, regularization methods, training time, etc.), the data set, the features [7] and even the initialization value of the random number generator, because these changes can lead to better results. Therefore, after each optimization, the process of data preparation, algorithm creation and training, is followed again and this process is repeated until the desired performance is achieved. Also, at this stage, an error analysis is performed, that consists in verifying the incorrect predictions (classifications) in order to identify the causes of erroneous detections and finding solutions. The evaluation of the optimized algorithm is be performed on the simulated data set presented in a previous chapter.

The stage referring to the **signal analysis algorithm** consists in building the model (deep learning algorithm) that will be trained, evaluated and optimized. The following functions and objects available in the Keras library within Tensorflow were used:

- *Sequential*—adding a sequential model;
- *save_model*, *load_model*—saving or uploading a model;
- *Dropout*, *l2*—adding regularization methods to penalize models based on their complexity, favoring simpler models that generalize better; these methods also prevent overfitting [28, 29];
- *BatchNormalization*—in order to increase the performance and stability of the neural networks by normalizing the activations of the previous layer (batch or groups of training data normalization) [30];
- *InputLayer*—defining of the input layer (actually, the feature array, taking into account that the input neurons do not have an activation function);

Table 8.1 Architecture of the DNN, CNN and LeNet-5 algorithms

DNN	CNN	LeNet-5
Input (1039)	Input (1039)	Input (1039)
DC_1 (520), R	$Conv1D_1$ (1,3) × 64, s = 1, LR	$Conv1D_1$ (1,11) × 64, s = 1, R
DC_2 (520), R	$Conv1D_2$ (1,3) × 64, s = 1, LR	$MaxPooling1D_1$ (1,3), s = 2
DC_3 (520), R	MaxPooling1D (1,2), s = 2	$Conv1D_1$ (1,13) × 128, s = 1, R
Output (1), sigmoid	Flattening	$MaxPooling1D_2$ (1,3), s = 2
–	DC (100), LR	Flattening
–	Output (1), sigmoid	DC_1 (1024), R
–	–	DC_2 (512), R
–	–	Output (1), sigmoid

DC densely connected, *LR* Leaky ReLU, *R* ReLU, *s* stride

- *Dense*—defining a fully connected layer;
- *Conv1D*—defining a convolutional layer;
- *LSTM*, *ConvLSTM2D*—defining LSTM layers;
- *MaxPooling1D*, *GlobalAveragePooling1D*—performing MaxPooling and AveragePooling subsampling operations;
- *Flatten*—flattening the data array;
- *LeakyReLU*, *ReLU*—activation functions.

The models were created sequentially, by defining the layers that make up the neural networks. The functions used for their creation are presented in Appendix. The structure of the developed algorithms is described in the following tables. They present the structures of the developed algorithms and the rows show the successive layers that form the neural networks (Tables 8.1, 8.2, 8.3 and 8.4).

The numbers in brackets have different meanings as shown in the following cases:

- In the case of input, output, fully connected and LSTM layers: the number of neurons;
- In the case of convolutional, subsampling and residual block layers: dimensions of the kernel used;
- When using the dropout technique: the percentage of neurons which are dropped.

The activation functions, the number of kernels used in convolutional networks and their stride are also specified.

Most of the investigated algorithms are designed for image analysis. The feature array used can be considered a 1D signal, and for its application it was necessary to modify the algorithms by replacing the two-dimensional (2D) convolution functions with the 1D versions and redefining the shape of the input data, thus obtaining 1D variants of these algorithms.

In the case of the ResNet-34 residual network, a high dropout value (0.8) is used to prevent overfitting. LeNet-5 was modified from the original version by using the ReLU activation function instead of the hyperbolic tangent and the MaxPooling

Table 8.2 Architecture of the AlexNet and VGG-16 algorithms

AlexNet	VGG-16	
Input (1039)	Input (1039)	Conv1D$_{10}$ (1,3) × 128, s = 1, R
Conv1D$_1$ (1,11) × 32, s = 1, R	Conv1D$_1$ (1,3) × 16, s = 1, R	MaxPooling1D$_4$ (1,2), s = 2
MaxPooling1D$_1$ (1,3), s = 2	Conv1D$_2$ (1,3) × 16, s = 1, R	Conv1D$_{11}$ (1,3) × 256, s = 1, R
Conv1D$_2$ (1,5) × 64, s = 1, R	MaxPooling1D$_1$ (1,2), s = 2	Conv1D$_{12}$ (1,3) × 256, s = 1, R
MaxPooling1D$_2$ (1,3), s = 2	Conv1D$_3$ (1,3) × 32, s = 1, R	Conv1D$_{13}$ (1,3) × 256, s = 1, R
Conv1D$_3$ (1,3) × 128, s = 1, R	Conv1D$_4$ (1,3) × 32, s = 1, R	MaxPooling1D$_5$ (1,2), s = 2
Conv1D$_4$ (1,3) × 128, s = 1, R	MaxPooling1D$_2$ (1,2), s = 2	Flattening
Conv1D$_5$ (1,3) × 128, s = 1, R	Conv1D$_5$ (1,3) × 64, s = 1, R	DC$_1$ (1024), R
MaxPooling1D$_3$ (1,3), s = 2	Conv1D$_6$ (1,3) × 64, s = 1, R	DC$_2$ (512), R
Flattening	Conv1D$_7$ (1,3) × 64, s = 1, R	Output (1), sigmoid
DC$_1$ (1024), R	MaxPooling1D$_3$ (1,2), s = 2	-
DC$_2$ (512), R	Conv1D$_8$ (1,3) × 128, s = 1, R	-
Output (1), sigmoid	Conv1D$_9$ (1,3) × 128, s = 1, R	-

DC densely connected, *R* ReLU, *s* stride

Table 8.3 Architecture of the ResNet-34 algorithm

ResNet-34	
Input (1039)	Res$_8$ (1,3) × 256, s = 2, R
Conv1D$_1$ (1,7) × 64, s = 2, R	Res$_9$ (1,3) × 256, s = 1, R
Dropout (0,8)	Res$_{10}$ (1,3) × 256, s = 1, R
Batch normalization	Res$_{11}$ (1,3) × 256, s = 1, R
MaxPooling1D (1,3), s = 2	Res$_{12}$ (1,3) × 256, s = 1, R
Res$_1$ (1,3) × 64, s = 1, R	Res$_{13}$ (1,3) × 256, s = 1, R
Res$_2$ (1,3) × 64, s = 1, R	Res$_{14}$ (1,3) × 512, s = 2, R
Res$_3$ (1,3) × 64, s = 1, R	Res$_{15}$ (1,3) × 512, s = 1, R
Res$_4$ (1,3) × 128, s = 2, R	Res$_{16}$ (1,3) × 512, s = 1, R
Res$_5$ (1,3) × 128, s = 1, R	GlobalAveragePooling1D
Res$_6$ (1,3) × 128, s = 1, R	Flattening
Res$_7$ (1,3) × 128, s = 1, R	Output (1), sigmoid

R ReLU, *Res* residual block, *s* stride

Table 8.4 Architecture of the LSTM, CNN LSTM and ConvLSTM algorithms

LSTM	CNN LSTM	ConvLSTM
Input (1039)	Input (258)	Input (258)
LSTM (100)	Conv1D$_1$ (1,3) × 64, s = 1, R	ConvLSTM2D (1,3) × 64, s = 1, R
DC (100), R	Conv1D$_2$ (1,3) × 64, s = 1, R	Flattening
Output (1), sigmoid	MaxPooling1D (1,2), s = 2	DC (100), R
–	Flattening	Output (1), sigmoid
–	LSTM (100)	–
–	DC (100), R	–
–	Output (1), sigmoid	–

DC densely connected, *R* ReLU, *s* stride

function instead of the AveragePooling, with the aim of increasing the algorithm's convergence speed and due to the wider use of the MaxPooling function [31]. The recurrent networks perform data processing over time and, therefore, the feature array was divided into 4 successive sequences, resulting in the smaller size of the input layer. Also, in order to be able to divide exactly the array into 4 sub-arrays (sequences), the first 3 samples of the spectral diagram were neglected, considering this an insignificant loss of information.

The batch normalization used in ResNet-34 is placed after the activation function to obtain a better performance [32]. The subsampling function (MaxPooling) prevents overfitting and reduces the amount of information to be processed (feature maps and number of hyperparameters). In addition, in the case of images, it offers the advantage of ensuring invariance to small translations [33], distortions or different light conditions. The ReLU activation function ensures nonlinearity, it is applied after each densely connected or convolutional layer and has been chosen to increase the convergence rate [34] and improve the classification performance [35]. The LeakyReLU function was used for the CNN algorithm, because it allows to obtain a better performance. Unlike ReLU, LeakyReLU allows a small gradient when the neuron is not active.

In all cases, the activation function of the last neuron is sigmoid and was chosen because it is recommended for binary classification applications [36], such as NQR detection. A single output neuron was chosen instead of two neurons, to reduce the complexity of the calculation and reduce the training time.

Table 8.5 shows the number of parameters used by each model to provide an overview of their complexity.

The trainable parameters are those whose values change during training. As seen, the LeNet-5 network has the highest number of parameters, followed by AlexNet.

At the end of the workflow, an optimized algorithm is obtained, with an acceptable detection performance for this application. During this stage, the algorithm will be

Table 8.5 The number of parameters of the algorithms developed

Algorithm	Total number of parameters	Total number of trainable parameters
DNN	1,083,161	1,083,161
CNN	3,321,609	3,321,609
LeNet-5	33,401,729	33,401,729
AlexNet	16,782,145	16,782,145
VGG-16	7,994,065	7,994,065
ResNet-34	7,235,777	7,218,753
LSTM	51,001	51,001
CNN LSTM	3,314,409	3,314,409
ConvLSTM	1,688,777	1,688,777

implemented on the NQR spectrometer and its performance will be monitored for further improvement by continuing to training with new data that will be available.

8.3 Training and Evaluation of the Signal Analysis Algorithms

This stage aims to train algorithms on real data and analyze the results obtained in order to choose the best solution. Appendix presents the learning curves for the developed algorithms. The detection accuracy and the value of the cost function are evaluated on both the training and test data sets, the latter representing 20% of the first set. The batch size used during training was calculated as the total number of acquisitions divided by 10, obtaining 2000 acquisitions. Thus, the use of large batches that could lead to the degradation of the model's ability to generalize was avoided [37].

In the case of VGG-16, only the samples of the spectral diagram were used as features, because the algorithm did not converge if the other features were included. Similarly, the training of the ResNet-34, LSTM and CNN LSTM networks with only the spectral diagram allowed to achieve better performance. In the case of LSTM, the use of the training set without scaling has led to improved accuracy and avoided peaks in the learning curves. The number of hidden layers and neurons was established empirically for DNN, through tests that assessed the accuracy of detection. This solution was chosen because there is no rule for choosing these parameters. In the case of DNN, the tests have indicated that scaling increases the convergence rate without affecting accuracy.

Algorithms have different convergence rates, the fastest being DNN, AlexNet, CNN LSTM and ConvLSTM. The number of training epochs was chosen empirically for each algorithm, taking into account the evolution of learning curves and training duration. The algorithms with the shortest training times are DNN, CNN, LSTM, while the slowest is ResNet-34.

Depending on the evolution of the learning curves, the following three situations can be encountered:

- Overfitting—the accuracy obtained with the training set is higher than that using test set, meaning that the model is not general because it has learned the training set too well and can no longer discriminate data different from it;
- Underfitting—the accuracy obtained with the training set is lower than that using the test set, meaning that the model cannot learn the training set and cannot generalize;
- Normal—the accuracy obtained with the training set is close to that obtained using the test set.

According to [7], these situations can be explained in the terms of bias and variance; the first can be considered as the error rate on the training set, and the second on the test set. Thus, the situation of overfitting occurs when the bias is low and the variance is high, and that of underfitting when the bias is high and the variance is low. In the case of the developed algorithms, with the exception of ResNet-34, the two learning curves are close, indicating normal training situations. In the case of ResNet-34, there is a strong overfitting throughout training, although a high dropout value was used. This can be explained by the fact that ResNet-34 is a very powerful algorithm, designed to classify complex signals (images), which learns very quickly the relatively simple signal (feature array) used in NQR detection.

After training the models, their evaluation was performed on the test set. It is mentioned that this represents the data set indicated in Fig. 8.11 and not the one used

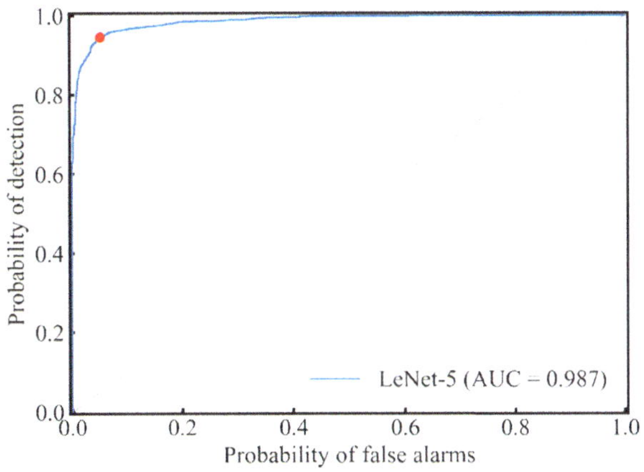

Fig. 8.11 ROC curve for the LeNet-5 algorithm (*AUC* Area Under Curve). Reprinted from Expert Sys. with Appl., vol. 182, C. Monea, Nuclear quadrupole resonance response detection using deep neural networks, 115,227, Copyright 2021, with permission from Elsevier [Or Applicable Society Copyright Owner]

during training (mentioned above). The evaluation consists in calculating the curve and measuring the accuracy of the model. The ROC curve allows the identification of the optimal detection threshold, which maximizes P_d and minimizes P_f. This threshold is used to assess the detection accuracy. Thus, the detection thresholds are established, and for each of them the true positive and false positive rates are determined by evaluating the entire test set. The ROC curve was calculated using the *roc_curve* function from the Scikit-learn library. The values obtained are illustrated in Fig. 8.11.

The optimal detection threshold is the point closest to the upper left corner of the plot and is represented by a red dot. Figure 8.12 also presents the AUC value, which shows the model's capacity to discriminate the detection classes; this value is desired to be 1. In the case exemplified for LeNet-5, the optimal threshold was identified as 0.66.

Following the determination of the optimal thresholds for all algorithms, their evaluation on the test set provided the results presented in Table 8.6. As seen, all algorithms have high and close accuracies. Despite the overfitting, ResNet-34 has

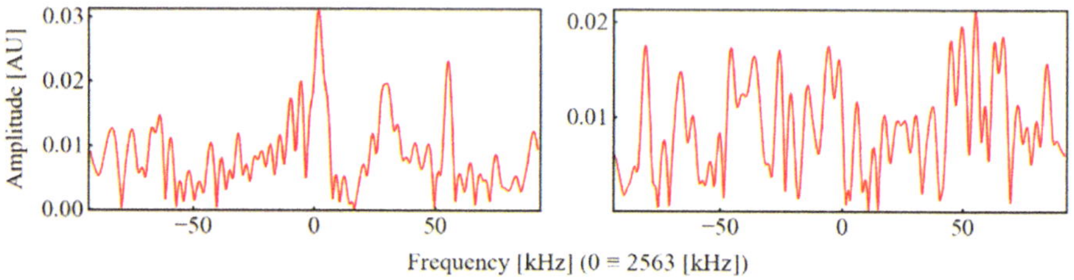

Fig. 8.12 Spectral diagram of the acquisitions 16,560 (left) and 16,559 (right), in the scenario NS paracetamol (*AU* Arbitrary units)

Table 8.6 Detection accuracy for the developed algorithms

Algorithm	Accuracy (%)
DNN	99.6
CNN	94.7
LeNet-5	94.9
AlexNet	**99.8**
VGG-16	99.3
ResNet-34	93.7
LSTM	99.3
CNN LSTM	99.6
ConvLSTM	99.6

Reprinted from Expert Sys. with Appl., vol. 182, C. Monea, Nuclear quadrupole resonance response detection using deep neural networks, 115,227, Copyright 2021, with permission from Elsevier [Or Applicable Society Copyright Owner]

achieved good performance, but, although it is the most complex algorithm developed, its accuracy is the lowest. DNN provided the second highest accuracy and is a simpler algorithm than the others. This indicates that a simpler algorithm is recommended for this type of detection. Generally, the more complex the signal, the more complex the algorithm used and the larger the data set [38]. It is noted that the recurrent networks have provided very good results.

According to the analysis performed in this stage, the AlexNet algorithm achieves the highest detection accuracy, of 99.8%, and is therefore proposed for optimization. The use of a more complex algorithm, such as AlexNet, to analyze a relatively simple signal can provide robustness to variations in noise or useful signal from the data set used.

8.4 Optimization of the Proposed Algorithm

The optimization is applied to increase the accuracy of the proposed algorithm as much as possible; it is desirable to reach the value of 100%. The first step of optimization is errors (incorrect classifications) analysis in order to identify their causes. The errors percentage in the case of the AlexNet model is 0.2%, representing 4 false negative and 4 false positive detections.

Table 8.7 indicates the records for which incorrect predictions (classifications) were obtained, presenting their index within the test set, the scanning scenario, the result of the classification from the point of view of the algorithm (detected/not

Table 8.7 Error analysis

Index in test set	Scenario	Substance response	ND	D	R	Cause
117	Paracetamol NS	Absent		✔	FP	RFI
1049	Paracetamol PS at 3 cm	Present	✔		FN	Spectrometer detection limit
1555	Sand NS	Absent		✔	FP	High noise components
1592	Paracetamol PS at 3 cm	Present	✔		FN	Spectrometer detection limit
2320	Paracetamol NS	Absent		✔	FP	RFI
2679	Sand NS	Absent		✔	FP	High noise components
3134	Paracetamol PS at 3 cm	Present	✔		FN	Spectrometer detection limit
3373	Sodium nitrite PS at 5 cm	Present	✔		FN	Spectrometer detection limit
% of total			50	50		

ND not detected, *D* detected, *R* result, *FP* False positive, *FN* False negative

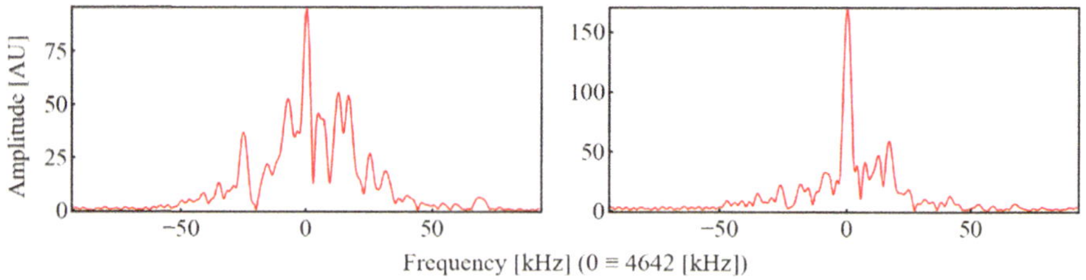

Fig. 8.13 Spectral diagram of the acquisitions 14,556 (left) and 14,555 (right), in the scenario NS sand (*AU* Arbitrary units)

detected) and the operational result (false positive/false negative), and the cause of the error, respectively. The index is different from the acquisition in the original data set, because the test set is a random subset of it. It is observed that the number of incorrect classifications is equal. The false positive results are obtained in the case of sand and paracetamol.

In order to highlight the causes of the errors, the following figures show the spectral diagrams of the signals obtained in some of these cases together with those resulting from previous acquisitions (Fig. 8.13).

Given that all the features are derived from the spectral diagram, then its analysis can indicate the causes of the errors. In the NS sand scenario, the resonance frequency of sodium nitrite was used to perform the scans. The plots on the left are those of the incorrectly classified acquisitions, and those on the right are those of the previous acquisitions. In the NS paracetamol scenario, it is observed that the plot on the left contains a spectral peak that can indicate the substance's response, but it does not appear in the previous acquisition. The conclusion is that the peak is caused by external or internal RFI. In the NS sand scenario, the higher noise components around the false spectral peak generated by the inverse piezoelectric effect caused the algorithm to interpret the spectral diagram as being generated by the substance's response. However, in the previous acquisition, these components are smaller and the algorithm classified correctly. The optimization is needed to better discriminate these signals.

The false negative results are obtained in scenarios where the substance is far from the detection coil. Figure 8.14 shows the signals obtained in this scenario.

The acquisition is made at the detection limit of the spectrometer, as one ca see from the response signal and noise levels. In the previous acquisition, the noise had a slightly lower level, which allowed the algorithm to classify correctly.

In order to reduce the number of incorrect classifications, three directions can be followed: (1) optimize the algorithm for better signal discrimination, (2) improve the algorithm by adding a noise and RFI rejection solution, and (3) improve the hardware of the spectrometer. The last option is not the subject of this study and will therefore not be addressed, and the second direction will be addressed in a future chapter. The approach according to the first direction is presented in the following.

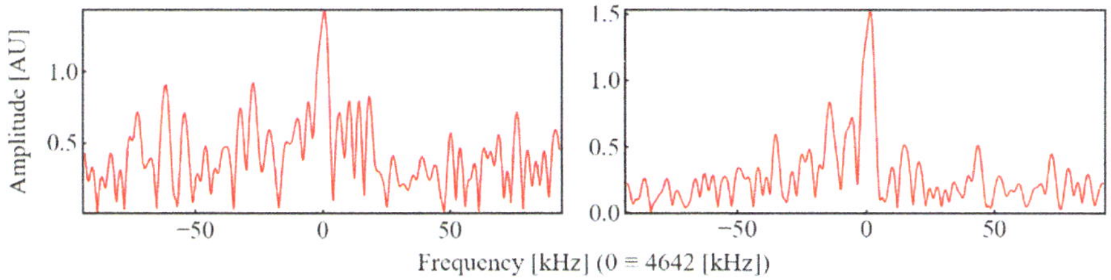

Fig. 8.14 Spectral diagram of the acquisitions 1018 (left) and 1017 (right), in the scenario sodium nitrite at 5 cm (*AU* Arbitrary units)

In order to investigate the impact of the different parameters on the classification accuracy, the *GridSearchCV* function from the Scikit-learn library was used, which compiles and trains the model for all combinations of the values provided, returning the scores obtained at the end. The following parameters were taken into account: the number of training epochs, the size of the training group, the optimization algorithm, its learning rate, the weight initialization function, the activation function, the number of neurons in the last layers, the size of the filters, area size used by the MaxPooling1D function. The results obtained by varying the values of the investigated parameters can be found in Appendix.

The value that provides the highest score has been identified. Each result is obtained by two cross-validations of the data set, which use randomly selected training/test sets from it, and averages the results. However, the model compiled with these values reached an accuracy of 99.7%, lower than the initial value. This can be explained by the fact that these parameters were analyzed individually, and not correlated. In addition, an attempt was made to modify the network's architecture by adding convolutional and fully connected layers, as well as to change the number of neurons and filters, respectively, and a small increase in accuracy was observed, reaching the limit value of 99.8%. It has been observed that better results are obtained when the spectral diagram is not normalized, this being explained by the fact that the PS differ from the NS by the higher average amplitude. Less redundant feature sets were also investigated; these were obtained by principal component analysis using a simple autoencoder (based on artificial neurons and convolutional), but the results were worse, due to the reduced information resulted by using these techniques.

This optimization approach did not provide a better model. However, by successively training the initial network, a better model was obtained. This is explained by the use of better initialization values, which are chosen randomly and are not under the control of the simulation program. The model reached the accuracy 99.9%, having only 4 false positive and no false negative predictions, as seen in Fig. 8.16, which is also significant from an operational point of view. Hereinafter, it will be called the optimized model. The optimal detection threshold resulting from the ROC analysis is 0.056695405.

The incorrect classifications are represented by the 4 false positives from Table 8.7, all caused by the spectrometer detection limit. These are illustrated in Figs. 8.15. It is remarkable that the algorithm distinguished the interference signals from the

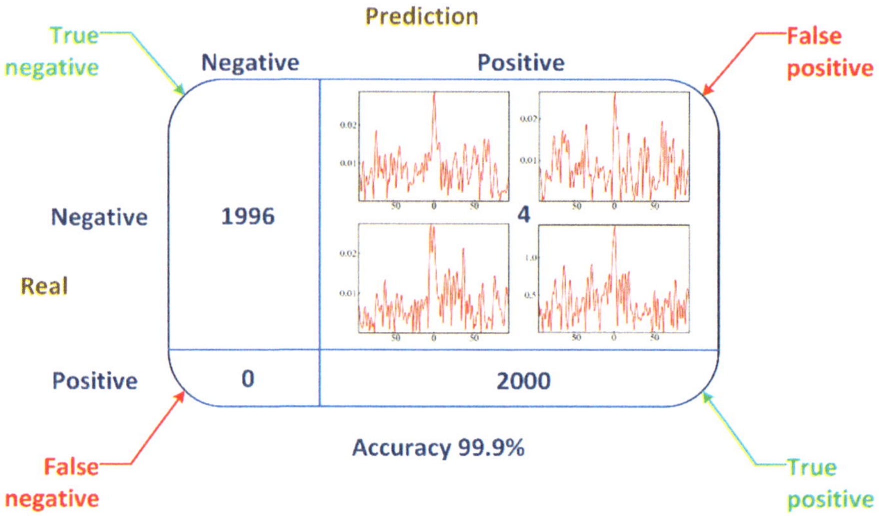

Fig. 8.15 The confusion matrix for the optimized algorithm. Reprinted from Expert Sys. with Appl., vol. 182, C. Monea, Nuclear quadrupole resonance response detection using deep neural networks, 115,227, Copyright 2021, with permission from Elsevier [Or Applicable Society Copyright Owner]

useful one, and eliminated all 4 false negatives given by the previous model. However, the model was verified on a single test set, which is a subset of the data set.

In order to obtain a more realistic result, a 10-fold cross-validation [39] was performed; thus, the data set is divided into 10 random subsets, 9 being used for training and 1 for testing, as illustrated in Fig. 8.16. The process is repeated 10 times, using in each round training and testing sets different from the initial ones, and at

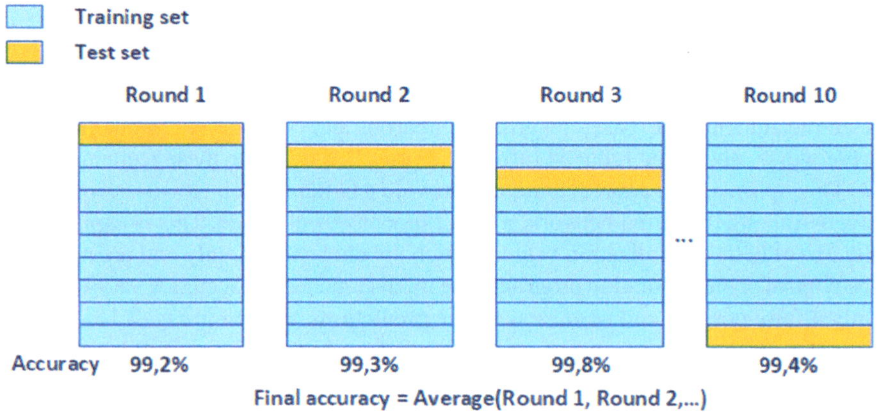

Fig. 8.16 Illustration of data sets used for cross-validation

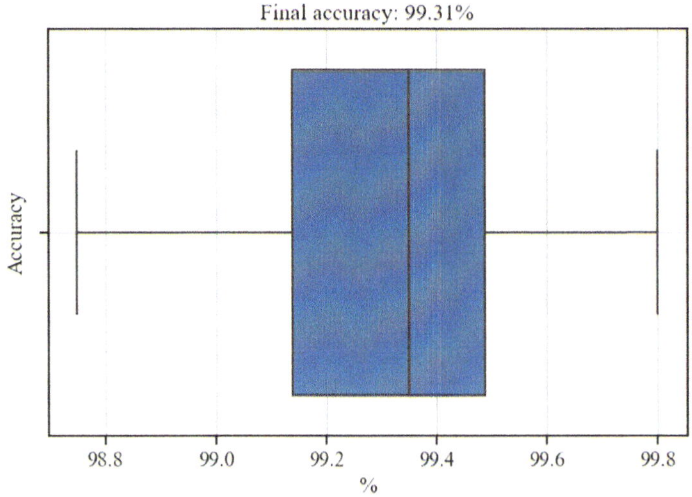

Fig. 8.17 Boxplot diagram of AlexNet model accuracy after 10 cross-validations

the end the results are averaged. In order to make a correct assessment, each set is normalized separately from the others. The data is also divided equally.

Figure 8.17 presents the result of cross-validation for the optimized AlexNet model.

It is observed that the values obtained are located in the range of 99.2–99.4%, with two extremes, and the average obtained is 99.31%, a good value that allows the use of the model in the operational (real) environment [40]. The average accuracy is lower than the previously reported value because, according to the cross-validation method, other training and test sets have been used than those utilized so far. Also, 10 validations were performed and, therefore, the training and test sets were composed of 18,000 acquisitions and 2000 acquisitions, respectively, with less records than initially used in the case of the test set.

The model was compared with the ensemble of classifiers (EC), identified in a previous chapter as the best performing algorithm. The simulated data sets were used, and in each scenario the training and evaluation of the AlexNet network was performed. The results are presented in Figs. 8.18 and 8.19.

The measurement of the classification accuracy was chosen because it is the most relevant performance indicator for this detection application. It is observed that in the scenario where the temperature substance is not different from the one corresponding to the reference frequency, the EC and AlexNet models have similar performances, slightly better in the case of the latter. In case of a temperature difference, AlexNet is superior at lower SNRs. In both scenarios, the algorithms classify perfectly at high SNR. It should be noted that the data sets at each SNR are relatively small, because they are composed of 1000 simulated acquisitions, and there is the possibility that the algorithms will have better results if the training sets are larger.

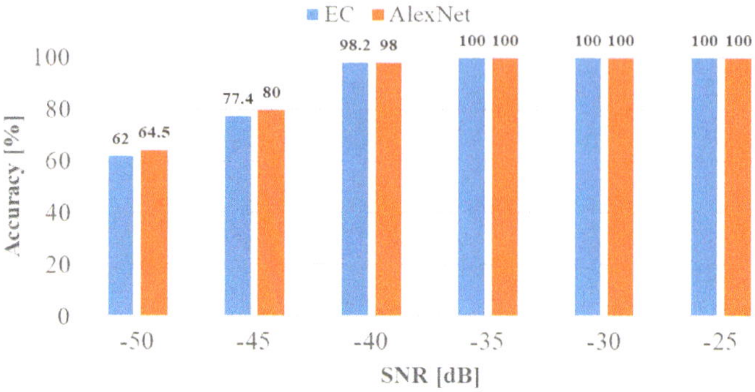

Fig. 8.18 Detection accuracy vs. SNR, in the case of simulated data, for EC and AlexNet

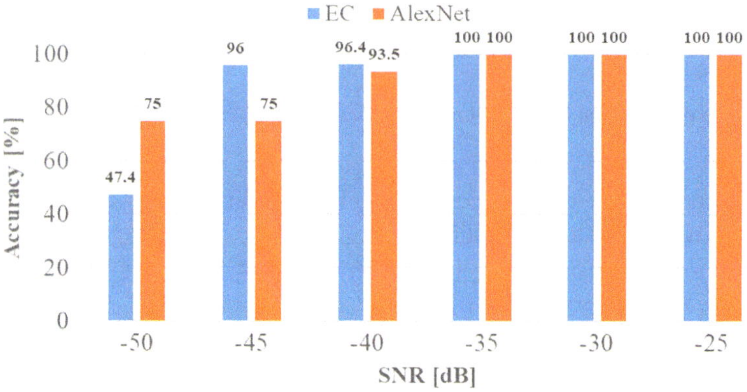

Fig. 8.19 Detection accuracy vs. SNR, in the case of simulated data and a temperature difference of 10 K, for EC and AlexNet

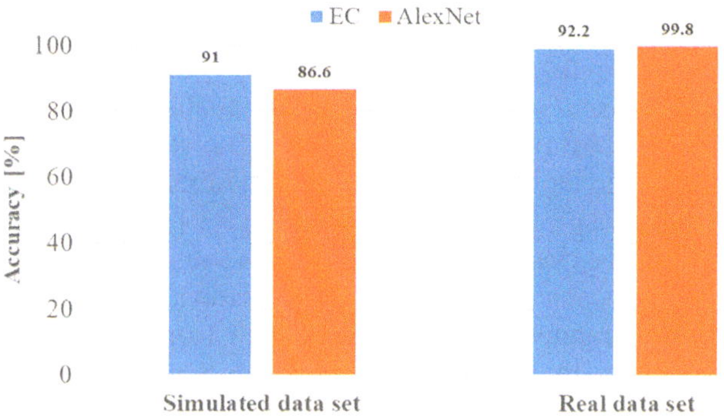

Fig. 8.20 Detection accuracy for simulated and real data sets, in the case of EC and AlexNet

Figure 8.20 presents the accuracy obtained after evaluating the algorithms on the entire simulated and real data sets, respectively.

In the case of simulated data, EC has an accuracy about 4% higher than AlexNet. But AlexNet is superior in the case of real data. The result in the case of simulated data can be explained by the fact that this set is more predictable and AlexNet is suitable for more complex signals, as presented in a previous chapter (signals containing nonlinearities due to the spectrometer, internal and external RFI, respectively, which were not simulated).

In conclusion, the optimized AlexNet network is able to discriminate the response signal in ambient noise conditions, including internal and external RFI, different temperature, false spectral peaks due to detuning or sand and even at the spectrometer detection limit. By visually analyzing the spectral diagrams of the acquisitions that the algorithm correctly classified, one can say that the optimized AlexNet model exceeds the human performance regarding the analysis of signals for NQR detection.

8.5 Conclusions

The chapter aimed to present some contributions in the development of signal analysis algorithms based on machine learning in order to improve the detection of substances by the nuclear quadrupole resonance technique.

The study started from the high-performance deep learning algorithms, modified them for the detection of NQR response signals and performed a comparative evaluation in order to choose the best solution. A total of 9 algorithms was selected and implemented in a dedicated programming environment, as follows: DNN, CNN, LeNet-5, AlexNet, VGG-16, ResNet-34, LSTM, CNN LSTM and ConvLSTM. The workflow and its implementation details were described. The architectures of the DNN and CNN models were proposed in this study, and the others were taken from the scientific literature. Also, certain algorithms, which were designed for image analysis, have been adapted to classify the one-dimensional signal in the case of NQR detection. The comparative analysis aimed at evaluating the classification accuracy, which is considered the most relevant performance indicator in this detection application. This showed that more complex algorithms, such as VGG-16 and ResNet-34, have lower performance, and simpler algorithms are more adequate. The best performing algorithm was AlexNet, reaching the accuracy of 99.8%, and was therefore proposed for optimization.

The optimization of the proposed algorithm aimed at obtaining the best classification performance and comparing it with the results of the previous study. The effect of varying the different models' parameters has on the detection accuracy was measured, such as the number of training epochs, the activation function or the optimization algorithm; also, their optimal values were identified. The model created with these values did not reach the initial performance, because the parameters were analyzed individually, and not correlated. However, a better model was obtained by successive training, which is explained by the use of better initialization values, which are chosen randomly and are not under the control of the simulation program. The optimized model also achieved a significant performance from an operational point of view, with an accuracy of 99.9%, yielding only 4 false positive and none false negative detections. Following the 10-fold cross-validation, the average accuracy of 99.31% was obtained, allowing the use of the model in the operational (real) environment. Also, the performance of the best algorithm identified in the previous study, EC, was evaluated on the real data set. Both classifiers have similar performance, AlexNet being superior at lower signal-to-noise ratios. In the case of simulated data, it reached a lower accuracy, because this data set is more predictable, and AlexNet is suitable for more complex signals, containing nonlinearities due to the spectrometer, internal and external RFI, respectively, which were not simulated.

Appendix

Source Code of the Functions Used to Generate the Deep Learning Models

1. **DNN model**

```
def BuildModel(nInput, nHidden, nOutput):
global DROPOUT, useBatchNormalization, useDropout

model = Sequential()

model.add(InputLayer(input_shape=(nInput,)))
model.add(Dense(nHidden, activation='relu'))

if (useDropout):
   model.add(Dropout(rate=DROPOUT))
if (useBatchNormalization):
model.add(BatchNormalization())
model.add(Dense(nHidden, activation='relu'))

if (useDropout):
   model.add(Dropout(rate=DROPOUT))
if (useBatchNormalization):
model.add(BatchNormalization())
model.add(Dense(nHidden, activation='relu'))
if (useDropout):
   model.add(Dropout(rate=DROPOUT))
if (useBatchNormalization):
   model.add(BatchNormalization())
model.add(Dense(nOutput, activation='sigmoid'))
# Compilare
model.summary()
model.compile(loss='binary_crossentropy',  optimizer=OPTIMIZER,
metrics=['accuracy'])

return model
```

2. CNN model

```
def BuildModel(nInput, nFinalDense, nOutput):
global DROPOUT, useBatchNormalization, useDropout

model = Sequential()
model.add(Conv1D(filters=64, kernel_size=3, activa-
tion=LeakyReLU(), input_shape=(nInput,1)))
model.add(Conv1D(filters=64, kernel_size=3, activa-
tion=LeakyReLU()))
if (useDropout):
   model.add(Dropout(rate=DROPOUT))
if (useBatchNormalization):
   model.add(BatchNormalization())
model.add(MaxPooling1D(pool_size=2))
model.add(Flatten())
model.add(Dense(nFinalDense, activation=LeakyReLU()))
model.add(Dense(nOutput, activation='sigmoid'))
# Compilare
model.summary()
model.compile(loss='binary_crossentropy', optimizer=OPTIMIZER,
metrics=['accuracy'])

return model
```

3. LeNet-5 model

```
def BuildModel(nInput, nOutput):
global DROPOUT, useBatchNormalization, useDropout

model = Sequential()

model.add(Conv1D(filters=64, kernel_size=11, strides=1, activa-
tion=ReLU(), input_shape=(nInput,1)))
model.add(MaxPooling1D(pool_size=3, strides=2))

model.add(Conv1D(filters=128, kernel_size=13, strides=1, activa-
tion=ReLU()))
model.add(MaxPooling1D(pool_size=3, strides=2))

model.add(Flatten())
model.add(Dense(1024, activation=ReLU()))
model.add(Dense(512, activation=ReLU()))
model.add(Dense(nOutput, activation='sigmoid'))
# Compilare
model.summary()
model.compile(loss='binary_crossentropy', optimizer=OPTIMIZER,
metrics=['accuracy'])

return model
```

4. AlexNet model

```
def BuildModel(nInput, nOutput):
global DROPOUT, useBatchNormalization, useDropout
```

```
model = Sequential()

model.add(Conv1D(filters=32, kernel_size=11, strides=1, activa-
tion=ReLU(), input_shape=(nInput,1)))
model.add(MaxPooling1D(pool_size=3, strides=2))

model.add(Conv1D(filters=64, kernel_size=5, strides=1, activa-
tion=ReLU()))
model.add(MaxPooling1D(pool_size=3, strides=2))
model.add(Conv1D(filters=128, kernel_size=3, strides=1, activa-
tion=ReLU()))

model.add(Conv1D(filters=128, kernel_size=3, strides=1, activa-
tion=ReLU()))
model.add(Conv1D(filters=128, kernel_size=3, strides=1, activa-
tion=ReLU()))
model.add(MaxPooling1D(pool_size=3, strides=2))

model.add(Flatten())
model.add(Dense(1024, activation=ReLU()))
if (useDropout):
   model.add(Dropout(rate=DROPOUT))
model.add(Dense(512, activation=ReLU()))
if (useDropout):
   model.add(Dropout(rate=DROPOUT))
model.add(Dense(nOutput, activation='sigmoid'))
# Compilare
model.summary()
model.compile(loss='binary_crossentropy', optimizer=OPTIMIZER,
metrics=['accuracy'])

return model
```

5. **VGG-16 model**

```
def BuildModel(nInput, nOutput):
global DROPOUT, useBatchNormalization, useDropout

model = Sequential()
model.add(Conv1D(filters=16, kernel_size=3, strides=1, activa-
tion=ReLU(), input_shape=(nInput,1)))
model.add(Conv1D(filters=16, kernel_size=3, strides=1, activa-
tion=ReLU()))
model.add(MaxPooling1D(pool_size=2, strides=2))

model.add(Conv1D(filters=32, kernel_size=3, strides=1, activa-
tion=ReLU()))
model.add(Conv1D(filters=32, kernel_size=3, strides=1, activa-
tion=ReLU()))
model.add(MaxPooling1D(pool_size=2, strides=2))

model.add(Conv1D(filters=64, kernel_size=3, strides=1, activa-
tion=ReLU()))
model.add(Conv1D(filters=64, kernel_size=3, strides=1, activa-
tion=ReLU()))
```

```
model.add(Conv1D(filters=64, kernel_size=3, strides=1, activa-
tion=ReLU()))
model.add(MaxPooling1D(pool_size=2, strides=2))

model.add(Conv1D(filters=128, kernel_size=3, strides=1, activa-
tion=ReLU()))
model.add(Conv1D(filters=128, kernel_size=3, strides=1, activa-
tion=ReLU()))
model.add(Conv1D(filters=128, kernel_size=3, strides=1, activa-
tion=ReLU()))
model.add(MaxPooling1D(pool_size=2, strides=2))

model.add(Conv1D(filters=256, kernel_size=3, strides=1, activa-
tion=ReLU()))
model.add(Conv1D(filters=256, kernel_size=3, strides=1, activa-
tion=ReLU()))
model.add(Conv1D(filters=256, kernel_size=3, strides=1, activa-
tion=ReLU()))
model.add(MaxPooling1D(pool_size=2, strides=2))

model.add(Flatten())
model.add(Dense(1024, activation=ReLU()))
model.add(Dense(512, activation=ReLU()))
model.add(Dense(nOutput, activation='sigmoid'))
# Compilare
model.summary()
model.compile(loss='binary_crossentropy', optimizer=OPTIMIZER,
metrics=['accuracy'])

return model
```

6. ResNet-34 model

```
DefaultConv1D = partial(Conv1D, kernel_size=3, strides=1,
padding="same", use_bias=False)

class ResidualUnit(Layer):
     def __init__(self, filters, strides=1, activation="relu",
**kwargs):

     super().__init__(**kwargs)

     self.activation = get(activation)
     self.main_layers = [
          DefaultConv1D(filters, strides=strides),
          Dropout(rate=DROPOUT),
          BatchNormalization(),
          self.activation,
         DefaultConv1D(filters),
         BatchNormalization()]
     self.skip_layers = []
     if strides > 1:
        self.skip_layers = [
         DefaultConv1D(filters, kernel_size=1, strides=strides),
          Dropout(rate=DROPOUT),
          BatchNormalization()]
```

```
    def call(self, inputs):
        Z = inputs
        for layer in self.main_layers:
            Z = layer(Z)
        skip_Z = inputs
        for layer in self.skip_layers:
            skip_Z = layer(skip_Z)
        return self.activation(Z + skip_Z)

    def get_config(self):

        config = super().get_config().copy()
        config.update({
            'activation': self.activation,
            'main_layers': self.main_layers,
            'skip_layers': self.skip_layers,
        })
        return config

def BuildModel(nInput, nOutput):
    global useDropout, DROPOUT

    model = Sequential()
        model.add(DefaultConv1D(64, kernel_size=7, strides=2,
input_shape=(nInput,1)))
    model.add(Dropout(rate=DROPOUT))
    model.add(BatchNormalization())
    model.add(Activation("relu"))
            model.add(MaxPooling1D(pool_size=3, strides=2,
padding="SAME"))
    prev_filters = 64
    for filters in [64] * 3 + [128] * 4 + [256] * 6 + [512] * 3:
        strides = 1 if filters == prev_filters else 2
        model.add(ResidualUnit(filters, strides=strides))
        prev_filters = filters
    model.add(GlobalAveragePooling1D())
    model.add(Flatten())
    model.add(Dense(nOutput, activation="sigmoid"))
    # Compilare
    model.summary()
            model.compile(loss='binary_crossentropy', opti-
mizer=OPTIMIZER, metrics=['accuracy'])

    return model
```

7. **LSTM model**

```
def BuildModel(nInput, nOutput):
global DROPOUT, useDropout

model = Sequential()
model.add(LSTM(100, input_shape=(nInput,1)))
if (useDropout is True):
    model.add(Dropout(rate=DROPOUT))
model.add(Dense(100, activation='relu'))
model.add(Dense(nOutput, activation='sigmoid'))
```

```
# Compilare
model.summary()
model.compile(loss='binary_crossentropy', optimizer=OPTIMIZER,
metrics=['accuracy'])

return model
```

8. CNN LSTM model

```
def BuildModel(nInput, nOutput):
global DROPOUT, useDropout

model = Sequential()
model.add(TimeDistributed(Conv1D(filters=64, kernel_size=3,
activation='relu'), input_shape=(None,nInput,1)))
model.add(TimeDistributed(Conv1D(filters=64, kernel_size=3,
activation='relu')))
if (useDropout is True):
   model.add(TimeDistributed(Dropout(rate=DROPOUT)))
model.add(TimeDistributed(MaxPooling1D(pool_size=2)))
model.add(TimeDistributed(Flatten()))
model.add(LSTM(100))
if (useDropout is True):
   model.add(Dropout(rate=DROPOUT))
model.add(Dense(100, activation='relu'))
model.add(Dense(nOutput, activation='sigmoid'))

# Compilare
model.summary()
model.compile(loss='binary_crossentropy', optimizer=OPTIMIZER,
metrics=['accuracy'])

return model
```

9. ConvLSTM model

```
def BuildModel(nInput, nSteps, nOutput):
global DROPOUT, useDropout

model = Sequential()
model.add(ConvLSTM2D(filters=64, kernel_size=(1,3), activa-
tion='relu', input_shape=(nSteps, 1, nInput, 1)))
if (useDropout is True):
   model.add(Dropout(rate=DROPOUT))
model.add(Flatten())
model.add(Dense(100, activation='relu'))
model.add(Dense(nOutput, activation='sigmoid'))

# Compilare
model.summary()
model.compile(loss='binary_crossentropy', optimizer=OPTIMIZER,
metrics=['accuracy'])

nnreturn model
```

Learning Curves for the Developed Algorithms

Accuracy and cost function for DNN (left) and CNN (right) algorithms.

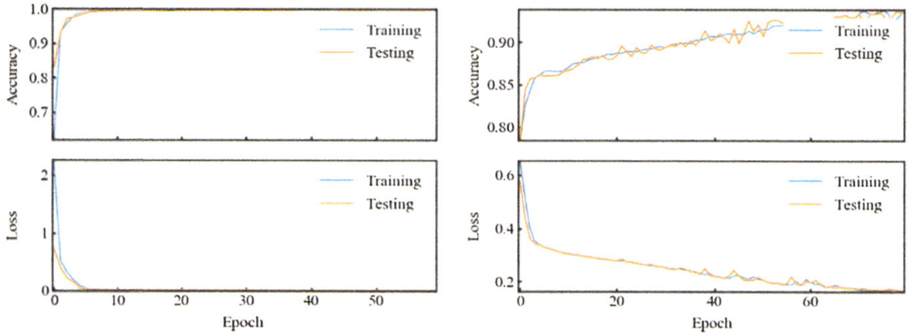

Accuracy and cost function for LeNet-5 (left) and AlexNet (right) algorithms.

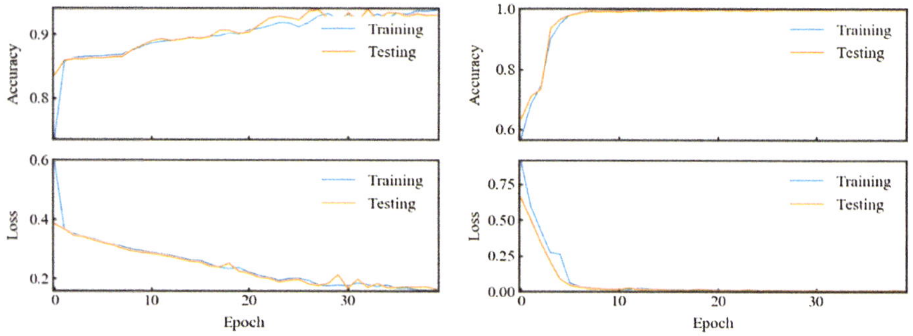

Accuracy and cost function for VGG-16 (left) and ResNet-34 (right) algorithms.

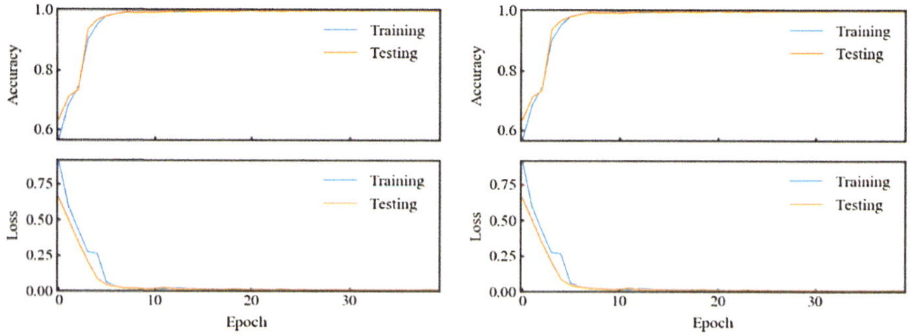

Accuracy and cost function for LSTM (left) and CNN LSTM (right) algorithms.

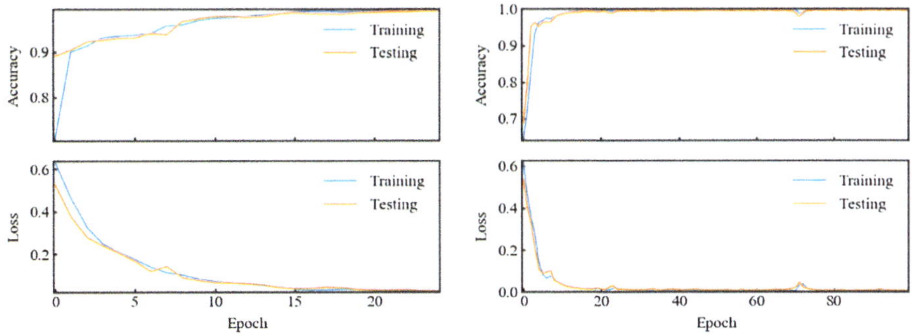

Accuracy and cost function for ConvLSTM algorithms.

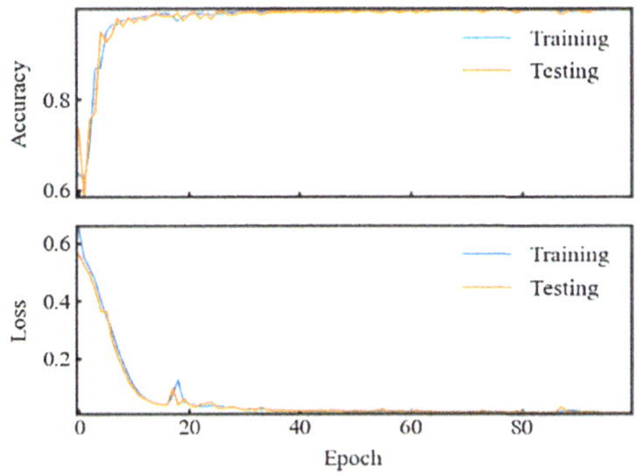

AlexNet Model Optimization Results

AlexNet model accuracy depending on the number of epochs and the batch size.

Number of epochs	10	40	80
Batch size			
100	0.993750 ± 0.001125	0.996563 ± 0.000187	0.996000 ± 0.000125
200	0.992125 ± 0.004375	0.992063 ± 0.002937	0.996437 ± 0.000437
400	0.994875 ± 0.000750	0.996312 ± 0.001063	0.996750 ± 0.000000
600	0.994813 ± 0.000813	**0.997313 ± 0.000688**	0.995500 ± 0.001500
800	0.990063 ± 0.000312	0.996437 ± 0.000437	0.996500 ± 0.000625

AlexNet model accuracy depending on the optimization algorithm, learning rate, weights initialization function and activation function.[1]

Optimization algorithm		Learning rate		Weights initialization function	
SGD	0.721437 ± 0.004187	0.0001	0.986437 ± 0.001312	Uniform	0.975375 ± 0.014500
RMSprop	0.990312 ± 0.000062	0.001	**0.990500 ± 0.002500**	lecun_uniform	**0.991563 ± 0.001687**
Adagrad	**0.992188 ± 0.000313**	0.01	0.498000 ± 0.000000	Normal	0.980500 ± 0.005625
Adadelta	0.777875 ± 0.030500	0.1	0.500000 ± 0.002000	Zero	0.498000 ± 0.000000
Adam	0.989312 ± 0.003063	0.2	0.498000 ± 0.000000	g_normal	0.991500 ± 0.000750
Adamax	0.988937 ± 0.001063	0.3	0.500000 ± 0.002000	g_uniform	0.984688 ± 0.000562
Nadam	0.898125 ± 0.093375	–	–	he_normal	0.988750 ± 0.000250
–	–	–	–	he_uniform	0.985687 ± 0.005938

AlexNet model accuracy depending on the activation function, kernel size and pool size.

Activation function		Kernel size		Pool size	
softmax	0.500000 ± 0.002000	3	0.976250 ± 0.013875	1	0.986125 ± 0.005375
softplus	0.727437 ± 0.002563	11	0.981937 ± 0.005312	2	0.985687 ± 0.006813
softsign	0.982313 ± 0.010188	15	**0.984687 ± 0.001687**	3	0.956875 ± 0.036500
relu	**0.991625 ± 0.001875**	20	0.981375 ± 0.005500	6	0.988250 ± 0.000000
tanh	0.977937 ± 0.007313	–	–	9	**0.989937 ± 0.000437**
sigmoid	0.500000 ± 0.002000	–	–	15	0.984375 ± 0.005375
hard_sigmoid	0.500000 ± 0.002000	–	–	20	0.988187 ± 0.001813
Linear	0.811750 ± 0.136250	–	–	–	–

References

1. Kaplan, A., Haenlein, M.: Siri, Siri, in my hand: Who's the fairest in the land? On the interpretations, illustrations, and implications of artificial intelligence. Bus. Horiz. **62**(1), 15–25 (2019)
2. Géron, A.: Hands-on Machine Learning with Scikit-Learn, Keras, and TensorFlow, 2nd edn. O'Reilly Media (2019)
3. Iana, G.V., Monea, C.: Coin recognition system based on a neural network. In: Proceedings of the 6th International Conference on Electronics, Computers and Artificial Intelligence, pp. 13–18 (2014)

[1] Reprinted from Expert Sys. with Appl., vol. 182, C. Monea, Nuclear quadrupole resonance response detection using deep neural networks, 115,227, Copyright 2021, with permission from Elsevier [Or Applicable Society Copyright Owner]

4. Monea, C., Iana, G.: Coin signal recognition based on neural networks. Sci. Bull. Univ. Pitesti **14**(1), 19–24 (2014)
5. Mohammed, M., Khan, M.B., Bashier, E.B.M.: Machine Learning: Algorithms and Applications, 1st edn. CRC Press, Boca Raton (2016)
6. Deng, L., Yu, D.: Deep learning: methods and applications. Found. Tr. Signal Process. **7**(3–4), 1–199 (2014)
7. Ng, A.: Machine Learning Yearning (2018)
8. Lazebnik, L.: Convolutional Neural Network Architectures: From LeNet to ResNet. University of Illinois, 13 Aug 2020. Internet: http://slazebni.cs.illinois.edu/spring17/lec01_cnn_architectures.pdf
9. Hu, J., Shen, L., Albanie, S., Sun, G., Wu, E.: Squeeze-and-excitation networks. In: IEEE/CVF Conference on Computer Vis. and Pattern Recognit (2018)
10. Brownlee, J.: Deep Learning for Time Series Forecasting (2018)
11. Schmidhuber, J.: Deep learning in neural networks: an overview. Neural Netw. **61**, 85–117 (2015)
12. Goodfellow, I., Bengio, Y., Courville, A.: Deep Learning. MIT Press (2016)
13. Cornell University: CS1114 Section 6: Convolution, 19 Aug 2020. Internet: https://www.cs.cornell.edu/courses/cs1114/2013sp/sections/S06_convolution.pdf
14. Ackermann, N.: Introduction to 1D Convolutional Neural Networks in Keras for Time Sequences, 19 Aug. 2020. Internet: https://blog.goodaudience.com/introduction-to-1d-convolutional-neural-networks-in-keras-for-time-sequences-3a7ff801a2cf
15. LeCun, Y., Bottou, L., Bengio, Y., Haffner, P.: Gradient-based learning applied to document recognition. Proc. IEEE **86**(11), 2278–2324 (1998)
16. Krizhevsky, A., Sutskever, I., Hinton, G.: ImageNet classification with deep convolutional neural networks. Comm. of the ACM **60**(6), 84–90 (2017)
17. Simonyan, K., Zisserman, A.: Very Deep Convolutional Networks for Large-Scale Image Recognition (2015). arXiv:1409.1556v6 [cs.CV]
18. He, K., Zhang, X., Ren, S., Sun, J.: Deep Residual Learning for Image Recognition (2015). arXiv:1512.03385 [cs.CV]
19. Sainath, T.N., Vinyals, O., Senior, A., Sak, H.: Convolutional, Long Short-Term Memory, Fully Connected Deep Neural Networks. In: IEEE International Conference on Acoustics, Speech and Signal Process (2015)
20. Shi, X., Chen, Z., Wang, H., Yeung, D.T., Wong, W.K., Woo, W.: Convolutional LSTM Network: A Machine Learning Approach for Precipitation Nowcasting (2015). arXiv:1506.04214 [cs.CV]
21. Wan, X.: Influence of feature scaling on convergence of gradient iterative algorithm. J. Phys.: Conf. Series **1213**, 032021 (2019)
22. Sarkar, D., Bali, R., Sharma, T.: Practical Machine Learning with Python: A Problem-Solver's Guide to Building Real-World Intelligent Systems, Apress (2017)
23. Scikit-Learn, MinMaxScaler, 19 Aug 2020. Internet: https://scikit-learn.org/stable/modules/generated/sklearn.preprocessing.MinMaxScaler.html
24. Hastie, T., Tibshirani, R., Friedman, J.: The Elements of Statistical Learning: Data Mining, Inference, and Prediction, 2nd edn. Springer, Berlin (2009)
25. Abu-Mostafa, Y.S., Magdon-Ismail, M., Lin, H.T.: Learning From Data. AMLBook (2012)
26. Kingma, D., Ba, J.: Adam: A Method for Stochastic Optimization (2017). arXiv:1412.6980 [cs.LG]
27. DiPietro, R.: A Friendly Introduction to Cross-Entropy Loss, 19 Aug 2020. Internet: https://rdipietro.github.io/friendly-intro-to-cross-entropy-loss/
28. Srivastava, N., Hinton, G., Krizhevsky, A., Sutskever, I., Salakhutdinov, R.: Dropout: a simple way to prevent neural networks from overfitting. J. Mach. Learn. Res. **15**, 1929–1958 (2014)
29. Bühlmann, P., Van De Geer, S.: Statistics for High-Dimensional Data. Springer, Berlin (2011)
30. Towards Data Science: Batch Normalization in Neural Networks, 19 Aug 2020. Internet: https://towardsdatascience.com/batch-normalization-in-neural-networks-1ac91516821c

31. Tsang, S.H.: Review: LeNet-1, LeNet-4, LeNet-5, Boosted LeNet-4 (Image Classification), 19 Aug 2020. Internet: https://medium.com/@sh.tsang/paper-brief-review-of-lenet-1-lenet-4-lenet-5-boosted-lenet-4-image-classification-1f5f809dbf17
32. Brownlee, J.: How to Accelerate Learning of Deep Neural Networks With Batch Normalization, 19 Aug 2020. Internet: https://machinelearningmastery.com/how-to-accelerate-learning-of-deep-neural-networks-with-batch-normalization/
33. Scherer, D., Müller, A., Behnke, S.: Evaluation of Pooling Operations in Convolutional Architectures for Object Recognition, Lect. Notes in Computer Sci., pp. 92–101 (2010)
34. Glorot, X., Bordes, A., Bengio, Y.: Deep Sparse Rectifier Neural Networks. In: Proceedings of the Fourteenth International Conference on Artificial Intelligence and Statistics, PMLR 15, pp. 315–323 (2011)
35. Jay Kuo, C.C.: Understanding Convolutional Neural Networks with A Mathematical Model (2016). arXiv:1609.04112 [cs.CV]
36. Nwankpa, C., Ijomah, W., Gachagan, A., Marshall, S.: Activation Functions: Comparison of trends in Practice and Research for Deep Learning. CoRR. abs/1811.03378 (2018)
37. Keskar, N.T., Mudigere, D., Nocedal, J., Smelyanskiy, M., Tang, P.T.P.: On Large-Batch Training for Deep Learning: Generalization Gap and Sharp Minima (2017). arXiv:1609.04836v2 [cs.LG]
38. Zhu, X., Vondrick, C., Fowlkes, C., Ramanan, D.: Do We Need More Training Data? (2015). arXiv:1503.01508v1 [cs.CV]
39. Bronshtein, A.: Train/Test Split and Cross Validation in Python, 19 Aug 2020. Internet: https://towardsdatascience.com/train-test-split-and-cross-validation-in-python-80b61beca4b6
40. Monea, C.: Nuclear quadrupole resonance response detection using deep neural networks. Exp. Syst. Appl. (2020) (in press)

Chapter 9
Solutions to Improve NQR Detection

9.1 Generalization of the Optimized Algorithm Using Transfer Learning

Transfer learning is an area of deep learning research focused on storing the knowledge gained by solving a problem and applying it to a similar problem [1]. Thus, the initially developed models can be generalized to other applications [2]. In the case of NQR detection, transfer learning can allow the generalization of the algorithm developed in a previous chapter to different detection equipment. The necessity of this approach is imposed by the fact that the existing NQR spectrometers may have different transfer characteristics and noise conditions, due to the components, electronic modules and their construction (shielding materials, module location). Thus, there is a possibility that the received signal may have spectral peaks due to hardware components, which differ from one equipment to another and may cause incorrect classifications. Another advantage of transfer learning is that it reduces the training time of the model if it has a complex architecture or the data volume is large.

The most commonly used approaches for applying transfer learning are the following: (1) developing a model and (2) using a pre-trained model. The first approach involves establishing the initial problem and acquiring a sufficient volume of data to be able to perform the modeling. After training the model on the initial data, it is used to solve the target problem, either by using parts of it or by adjusting it to the new data (fine tuning). The second approach is to take an already trained model and use it to solve the new problem, similar to the first case. Although there are several pre-trained models available for image classification [3], no solutions have been developed for NQR detection. Therefore, the first approach will be followed for the application transfer learning, as illustrated in Fig. 9.1.

C. Monea and N. Bizon, *Signal Processing and Analysis Techniques for Nuclear Quadrupole Resonance Spectroscopy*, Signals and Communication Technology,
https://doi.org/10.1007/978-3-030-87861-0_9

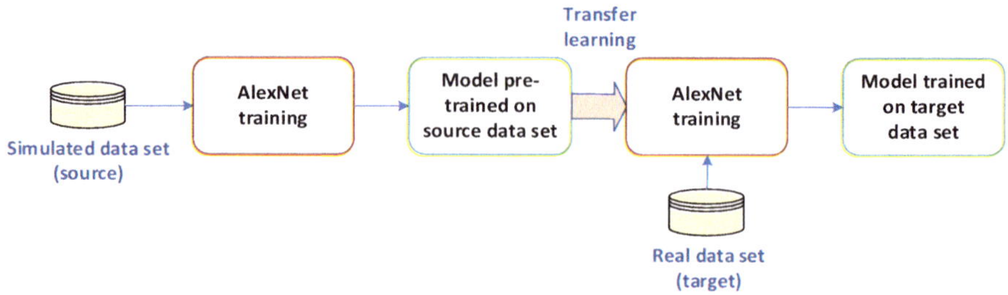

Fig. 9.1 Transfer learning applied in NQR detection

The AlexNet network is trained on the source data set, and the knowledge obtained is transferred using the data from the target spectrometer, on which the model will run. This requires the existence of at least two different spectrometers to acquire the source data and to improve the initial model with the data from the target equipment. Since there is only one spectrometer available, the simulated data set will be used to train the network, and the real one to improve it using transfer learning. Figure 9.2 shows the learning curves of the AlexNet network on the simulated data set.

The algorithm converges after about 25 training epochs, and the accuracy on the training and test (validation) sets is limited to around 85%. Following the evaluation of the model on the test set consisting of 2400 acquisitions, representing 20% of the 12,000 acquisitions that make up the simulated data set, the accuracy of 86.5% was obtained for the detection threshold 0.6377587. Also, there were 151 false positive and 171 false negative predictions, respectively.

Several strategies can be applied for the implementation of transfer learning, depending on the characteristics of the source and target data sets, as illustrated in Fig. 9.3.

The figure illustrates four cases in which the two data sets can be found depending on the dimensions and the similarity between them. If the target set is large and

Fig. 9.2 Accuracy and cost function for the AlexNet algorithm trained on the simulated data set

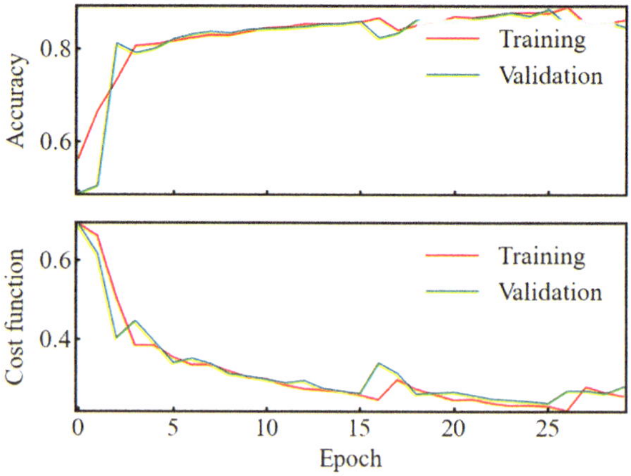

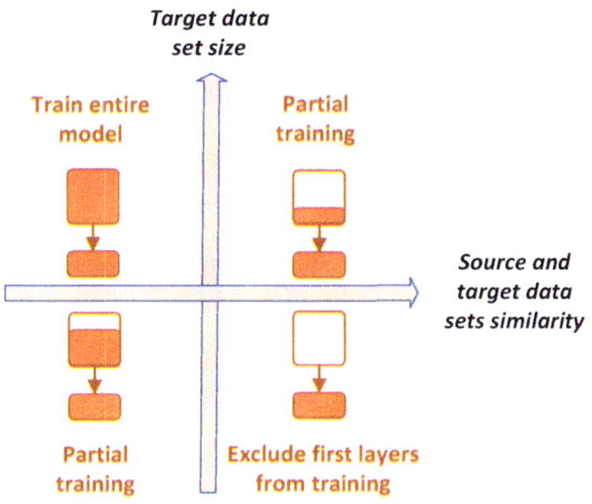

Fig. 9.3 Strategies for implementing transfer learning

different from the source set, then the strategy corresponding to quadrant 1 is applied, namely, training the entire model. However, if it is similar to the source, then any strategy can be applied, but it is recommended to train only certain layers (partial training). Similarly, when the target set is small and different from the source, the same strategy is applied, possibly with the training of a larger number of layers. In the latter case, it is recommended to train/modify only the last fully connected layers. Taking into account that the simulated and real data sets are similar and wishing to perform a study of the influence of the number of layers on the detection performance, the accuracy obtained in the case of training a different number of layers is investigated. Thus, for the implementation of transfer learning, the first convolutional layers will be frozen (blocked), and the training process will continue only for the next layers. Freezing means blocking the weights, representing the values of the filters from the convolutional layers and the actual weights from the fully connected ones, respectively. Thus, the first layers are used to extract the general features, and the last ones to identify the relationships between them and perform the classification. It also reduces the training time because a smaller number of layers is involved in this process. Freezing the layers is done by setting their *trainable* property accordingly.

Several strategies can be applied for the implementation of transfer learning, depending on the characteristics of the source and target data sets, as illustrated in Fig. 9.3.

The table also shows the shape of the data provided by each layer and the corresponding number of trainable parameters. In order to evaluate the influence that the number of frozen layers has on the network's learning capacity, first the whole network was trained (0 blocked layers), then, successively, more and more layers were frozen. Table 9.1 presents the results of this study, and the number of frozen layers indicates all the blocked layers up to the one indicated by this number.

The optimal detection threshold was obtained based on the ROC analysis, and the accuracy resulted from the evaluation of the model on the test set using this threshold. The confusion matrix is presented on different rows. It is observed that, in general,

Table 9.1 The influence that the number of frozen layers has on the network performance

Number of frozen layers	Optimal detection threshold	Accuracy (%)	True positives	True negatives	False positives	False negatives
0	0.46	99.8	1996	1997	3	4
1	0.62	99.7	1996	1994	6	4
2	0.53	99.7	1996	1995	5	4
5	0.22	99.8	1996	1997	3	4
10	0.69	93.9	1905	1854	146	95
11	0.51	93.9	1777	1889	111	223

the model learns better the target data set when fewer layers are frozen. Using a smaller number of trainable layers does not allow the network to learn the features of the target dataset. The last case represents, practically, the evaluation on real data of the initial network, trained on the simulated data set. The best model is obtained by freezing the first 3 layers and reaches an accuracy of 99.8%, providing the lowest number of false negatives. It was evaluated on the actual test set, reaching an accuracy of 99.85%, approximating that of the optimized model developed in the previous section. Following the re-evaluation on the simulated test set, it reached an accuracy of 85.08%, about one percentage point lower than the initial one. Also, tenfold cross-validation was performed on the two data sets and the following accuracies were obtained: 75.79% on the simulated data set and 99.18% on the real data set. The final values are lower in both situations and the better classification on real data is noted.

9.2 Noise Reduction Using Autoencoders

NQR detection is affected by unwanted signals represented by noise and radio frequency interference, as shown in a previous chapter. Thus, one idea to improve detection is to eliminate these signals. One of the machine learning solutions that can be applied in this situation is represented by autoencoders (AE).

The autoencoder is a particular case of a neural network that learns to copy the input data to its output. The network consists of an input layer, one or more hidden layers, and an output layer. The first half of the network has the role of encoding (representing) the input data in a smaller form, learning and generating their latent features. The second half aims to decode the data represented in latent form by reconstructing the initial data. The training is unsupervised, applying at the input the data to be learned, and at the output the desired data. Thus, among the first applications of AE is the reduction of data size; autoencoders are, in fact, non-linear variants of principal components analysis technique. They are also used for noise reduction (denoising) when noise-affected data is applied to the input and undisturbed data is used at the output. In this case, the AE learns to recover the initial data, reconstructing them without noise (thus, correcting the data affected by noise).

This study focuses on reducing disturbing signals using AE. Two different architectures are investigated: (1) autoencoder based on artificial neurons (DNN-AE) and (2) convolutional autoencoder (ConvAE). The functions for generating these models are presented in Appendix A.9.4.1. The spectral diagrams of the acquired signals are used for training the AE, and the resulting models are applied before the AlexNet network for denoising, as shown in Fig. 9.4.

Only the spectral diagrams are used to train the autoencoders, because the other features used by the AlexNet model are extracted from them. In order to perform the training, an additive Gaussian noise with zero mean, unit variance and the amplitude multiplied by the factor 0.5 is added to the initial signals. The noise-affected signals are applied at the input, and the undisturbed signals are applied at the output. Thus, the network will learn to reconstruct the signals without noise. Tables 9.2 and 9.3 present the structure of the two autoencoders.

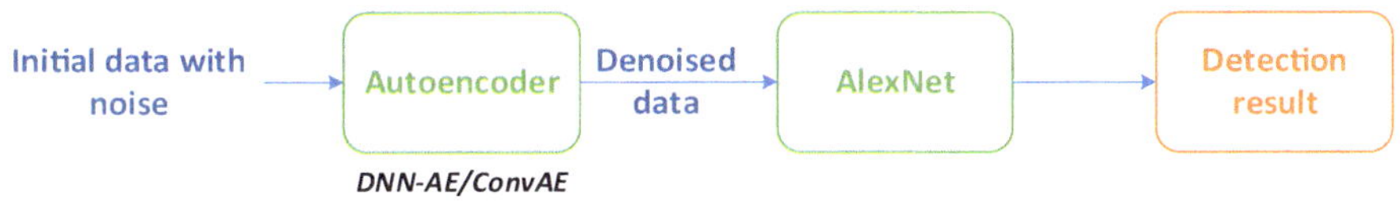

Fig. 9.4 Network structure consisting of the autoencoder and AlexNet networks

Table 9.2 Structure of the autoencoder with artificial neurons

Components	Layer description
Encoder	Input (1024)
	DC_1 (896), LR
	DC_2 (768), LR
	DC_3 (640), LR
	DC_4 (512), LR
	DC_5 (384), LR
	DC_6 (256), LR
	DC_7 (128), LR
Decoder	DC_8 (256), LR
	DC_9 (384), LR
	DC_{10} (512), LR
	DC_{11} (640), LR
	DC_{12} (768), LR
	DC_{13} (896), LR
	DC_{14} (1024), si

DC densely connected, *LR* Leaky ReLU, *si* sigmoid

Table 9.3 Structure of the convolutional autoencoder

Components	Layer description
Encoder	Input (1024)
	Conv1D_1 (1,3) × 64, s = 1, R
	MaxPooling1D_1 (1,2), s = 1
Decoder	Conv1D_2 (1,3) × 64, s = 1, R
	UpSampling_1 (2)
	Conv1D_3 (1,3) × 1, s = 1, si

si sigmoid, *s* stride

The different dimensions of the two AE are observed. The semi-linear activation function was selected because it ensures nonlinearity, decreases the convergence time [4] and increases the classification performance [5]. In the case of the convolutional AE, the *UpSampling* function from the Keras library was used to implement the decoder. It performs the oversampling operation by repeating the values twice, in this case (the factor is presented in brackets). The LeakyReLU activation function was chosen for the autoencoder with artificial neurons, because it provided better results in the tests performed.

Data normalization is done by dividing the spectral diagram's samples to the maximum value, before adding the noise, because it provided better results. The mean square error was chosen as a cost function, taking into account that no classification is made and it is desired to reconstruct the useful signal from the noise as accurately as possible. Figure 9.5 shows the learning curves for the two autoencoders.

A number of 70 training epochs were used. It is observed that the convolutional AE converges faster than the first, and both reach a low cost value. The following figures illustrate the spectral diagrams before and after the application of the convolutional AE. The plots show the initial spectral diagrams from PS and NS, those obtained after the application of Gaussian noise and the result after denoising using the ConvAE (Figs. 9.6 and 9.7).

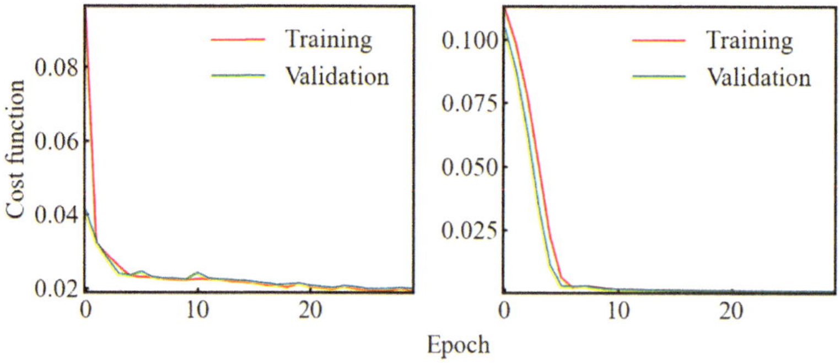

Fig. 9.5 Cost function for the artificial neuron-based (left) and convolutional (right) autoencoder

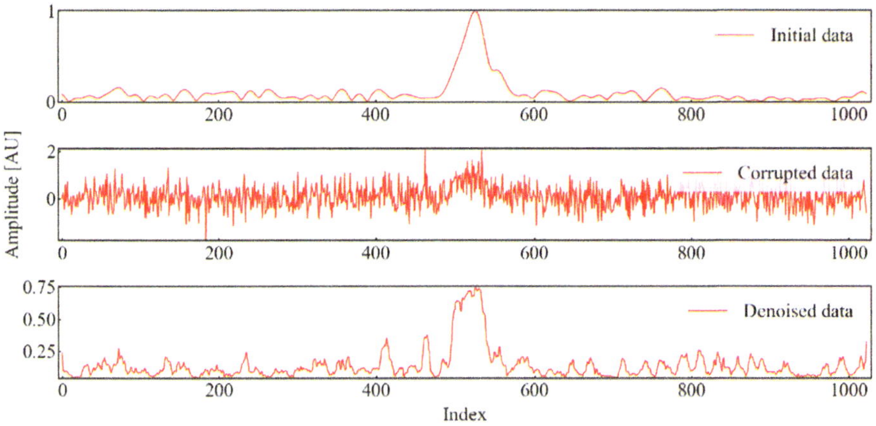

Fig. 9.6 Spectral diagrams without noise (top), with noise (center) and after denoising (bottom), in the presence of the NQR response signal

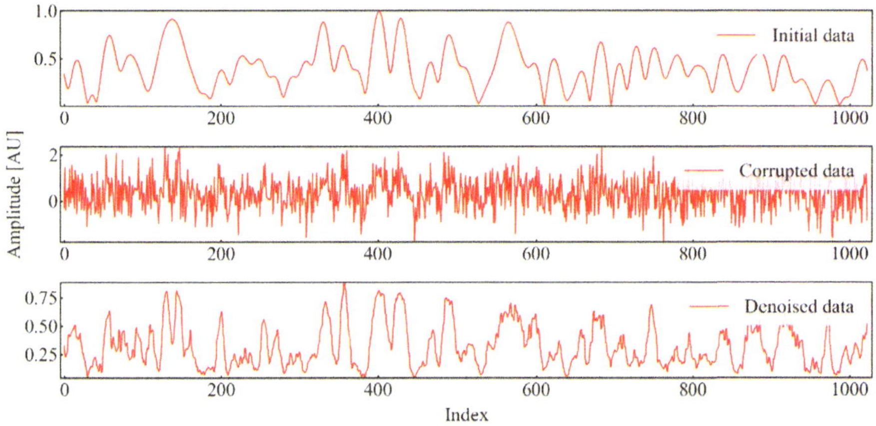

Fig. 9.7 Spectral diagrams without noise (top), with noise (center) and after denoising (bottom), in the absence of the NQR response signal

The model's ability to reconstruct the initial useful signal from noise, even when it is completely covered, is remarkable.

A study was performed at multiple signal-to-noise ratios to visualize the model's performance under different noise conditions. The noise is generated at specific SNRs by defining its variance in correlation with the desired SNR, as follows:

$$\sigma_w^2 = \frac{\sigma_s^2}{10^{\mathrm{SNR_{dB}}/10}} \tag{9.1}$$

Table. 9.4 Average accuracy of the AlexNet model combined with autoencoders, after tenfold cross-validation

Signal-to-noise ratio [dB]	AlexNet	AlexNet with DNN-AE	AlexNet with ConvAE
−30	59.47	70.61	81.73
−25	63.88	85.43	85.12
−20	82.44	92.08	90
−15	80.73	95.13	95.46
−10	89.23	97.61	97.88
−5	92.79	98.58	98.85
0	93.49	99.06	99.11
5	96.25	98.87	99.07
10	99.46	99.07	99.12
15	99.49	99.08	99.14
20	99.44	99.17	99.13

where SNR_{dB} is the desired value; the average noise is zero. The AlexNet model is evaluated on the real data set at different SNRs, for each autocoder. Initially, the noise training sets are generated at the following SNRs: −30, −25, −20, −15, −10, −5, 0, 5, 10, 15 and 20 dB. Then, the autoencoders are trained with these sets, and finally, the model is evaluated on the test set by performing a tenfold cross validation. Table 9.4 shows the average accuracy obtained for each SNR, in three cases: (1) using only the AlexNet model, (2) combining the DNN-AE model with the AlexNet network and (3) combining the ConvAE model with the AlexNet network.

The performance of the AlexNet model is very good at SNRs over 10 dB, but decreases rapidly to approx. 60% at −30 dB. As seen, the autoencoders significantly improve the detection, increasing the accuracy by approx. 10 and 20% at −30 dB, respectively. The increase is maintained throughout the investigated range. The AlexNet model combined with autoencoders converges faster than the initial network, reaching the maximum accuracy at 0 dB. The two autoencoders bring similar improvements and the better performance provided by the ConvAE model at −30 dB is noticeable. The difference in accuracy between the original AlexNet model and the one combined with AE is less than 0.5% at SNRs over 10 dB, and is thus negligible. Taking into account the better performance at lower SNRs, the ConvAE model is proposed as the best noise reduction solution for NQR detection. The advantages of using a deep learning solution are that it avoids the need to develop algorithms specific to certain types of noise and RFI, as presented in a previous chapter, and is based on DL mechanisms to identify and cancel any type of disturbance learned during the training process.

9.3 Conclusions

This chapter has proposed solutions to improve nuclear quadrupole resonance detection using the optimized AlexNet algorithm presented a previous chapter. Initially, the generalization of the algorithm using transfer learning was proposed, so that it achieves good performance on spectrometers different than the one used to make the acquisitions. This is necessary because other spectrometers may have different transfer characteristics, which may influence the response signal. The algorithm was trained on the simulated data set, and the knowledge obtained was transferred to the real one. The influence the number of frozen layers has on the detection performance was evaluated and it was shown that the model approximates the accuracy of the optimized one. Then, the application of autoencoder networks to reduce disturbing signals was investigated. These were simulated by adding Gaussian noise to the useful signal and the detection performance was evaluated at several signal-to-noise ratios in the range of [−30, 20] dB. The autoencoders have been shown to have a remarkable denoising ability even when the useful signal is completely covered, significantly improving the detection accuracy by up to 20% at −30 dB. This solution has the advantages of avoiding the need to develop algorithms specific to certain types of noise and interference, and is based on deep learning mechanisms to identify and eliminate any type of disturbance learned during the training process.

Appendix

Source Code of the Functions Used to Generate the Autoencoder Models

1. **Artificial neurons-based autoencoder model**

```
def BuildModel(nInput):
input_dim = Input(shape = (nInput, ))
# Encoder Layers
encoded1 = Dense(896, activation=LeakyReLU())(input_dim)
encoded2 = Dense(768, activation=LeakyReLU())(encoded1)
encoded3 = Dense(640, activation=LeakyReLU())(encoded2)
encoded4 = Dense(512, activation=LeakyReLU())(encoded3)
encoded5 = Dense(384, activation=LeakyReLU())(encoded4)
encoded6 = Dense(256, activation=LeakyReLU())(encoded5)
encoded7 = Dense(encoding_dim, activation=LeakyReLU())(encoded6)
# Decoder Layers
decoded1 = Dense(256, activation=LeakyReLU())(encoded7)
decoded2 = Dense(384, activation=LeakyReLU())(decoded1)
decoded3 = Dense(512, activation=LeakyReLU())(decoded2)
decoded4 = Dense(640, activation=LeakyReLU())(decoded3)
decoded5 = Dense(768, activation=LeakyReLU())(decoded4)
decoded6 = Dense(896, activation=LeakyReLU())(decoded5)
```

```
decoded7 = Dense(nInput, activation = 'sigmoid')(decoded6)
# Combine Encoder and Deocoder layers
model = Model(inputs = input_dim, outputs = decoded7)
model.summary()
# Compile the Model
model.compile(optimizer=OPTIMIZER, loss='mean_squared_error')
return model
```

2. **Convolutional autoencoder model**

```
def BuildModel(nInput):
input_layer = Input(shape=(nInput,1))
# encoder
h = Conv1D(64, 3, activation=ReLU(), padding='same')(input_layer)
h = MaxPooling1D(2, padding='same')(h)
# decoder
h = Conv1D(64, 3, activation=ReLU(), padding='same')(h)
h = UpSampling1D(2)(h)
output_layer = Conv1D(1, 3, activation='sigmoid',
padding='same')(h)
model = Model(input_layer, output_layer)
model.summary()
# Compile the Model
model.compile(optimizer=OPTIMIZER,
loss='mean_squared_error')#loss='binary_crossentropy')
return model
```

References

1. Pan, S.J., Yang, Q.: A survey on transfer learning. IEEE Trans. Knowl. Data Eng. **22**(10), 1345–1359 (2010)
2. Goodfellow, I., Bengio, Y., Courville, A.: Deep Learning. The MIT Press (2016)
3. Ngoc, A.H.: keras-yolo2, 19 Aug 2020. Internet: https://github.com/experiencor/keras-yolo2
4. Glorot, X., Bordes, A., Bengio, Y.: Deep sparse rectifier neural networks. In: Proceedings of the Fourteenth International Conference on Artificial Intelligence and Statistics, PMLR 15, pp. 315–323 (2011)
5. Jay Kuo, C.C.: Understanding Convolutional Neural Networks with a Mathematical Model (2016). arXiv:1609.04112 [cs.CV]

Chapter 10
Implementation of a Signal Pre-processing, Processing and Analysis System for Nuclear Quadrupole Resonance

10.1 Hardware Implementation of the Detection System

The system developed represents an NQR spectrometer that operates in pulses and is realized for the detection of the prohibited substances in envelopes, packages or small luggage. The system integrates multiple solutions to the factors that affect detection and for noise reduction, such as fiber optic transmission, during data acquisition, of the control signals to modules near the analog circuits, and powering low noise amplifiers (LNA) from isolated batteries.

The system uses an excitation sequence based on multiple pulses. The equipment is designed for scanning explosives, narcotics, toxic substances and drugs in the frequency band 0.4–6 MHz. It can be used at any security checkpoint where NQR scanning equipment is required, such as airports, customs, ports, post offices, parcel processing points, embassies, prisons, hotels, banks, companies, etc. Figure 10.1 shows the system block diagram.

The detector is composed of the transmission (orange) and reception (blue) chains. The detection control and signal processing block is responsible for generating the signals required during NQR excitation and acquisition (e.g., pulse sequence, control signals). It is based on a Spincore RadioProcessor module and an embedded system with a dsPIC33E microcontroller. The RF amplifier is responsible for amplifying the excitation signal in order to reach the levels required for NQR detection. It is a Tomco BT00250, with a peak power of 250 W. The LC resonant circuit is composed of two variable vacuum capacitors controlled by the tuning system. The probe is a planar coil on which the substance/item to be scanned is placed. The automatic tuning system is composed of two stepper motors, one for each capacitor, and is capable of coarse tuning to predefined positions (corresponding to the resonance frequencies of the substances in the database) and fine tuning around a specific frequency. The automatic tuning system is controlled via the embedded system. The reception chain is composed of two low noise amplifiers (LNAs), reaching a total gain of aprox. 98 dB. The detector is controlled by a software application with graphical user interface running on a touchscreen computer integrated into the system.

C. Monea and N. Bizon, *Signal Processing and Analysis Techniques for Nuclear Quadrupole Resonance Spectroscopy*, Signals and Communication Technology,
https://doi.org/10.1007/978-3-030-87861-0_10

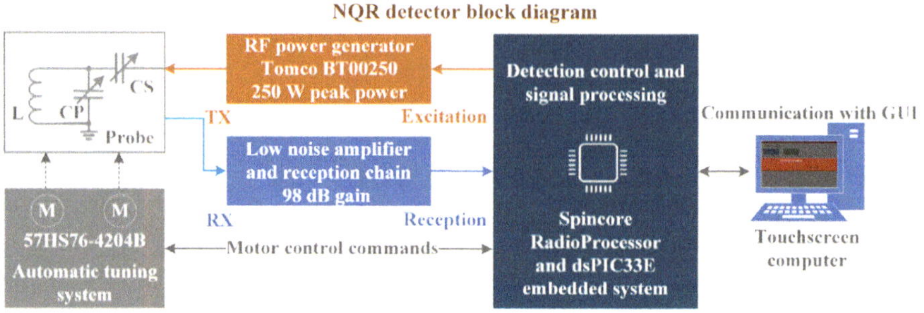

Fig. 10.1 General block diagram of the NQR detection system [1]. Reprinted from Monea et al. [1], Copyright 2020, with permission from Elsevier [Or Applicable Society Copyright Owner]

In order to compensate for the factors that affect the detection, several solutions were implemented. Firstly, the control signals from the detection control and signal processing module are sent via fiber optic cables to the noise sensitive circuits (e.g., transmission-reception switch and capacitor bank located near the resonant circuit). Secondly, in order to cover the entire frequency range, a capacitor bank is placed in parallel with the resonant circuit. The bank is composed of blocks of capacitors and includes a relay for each block; it provides an adjustable capacitance of up to 29.5 nF. The detection enclosure where the coil is placed has multiple layers of shielding composed of rubber, iron, copper and conductive felt. Also, its small size (35 × 35 × 40 cm) limits the maximum distance between the coil and the scanned item. The LNAs are powered from batteries which are electrically isolated from the rest of the detector in order to reduce acquisition noise; moreover, battery charging is disabled during signal acquisition. Also, the detector uses the WURST-QCPMG excitation sequence to compensate for the unknown substance temperature; this is estimated using an external temperature sensor which is used to set the reference excitation frequency.

A detailed description of the system's hardware can be found in [1].

10.2 Software Implementation of the Detection System

The software component of the NQR detection system is responsible for controlling its operation, monitoring it and to allowing its operation by the user. It consists of two software sub-components, as follows: (1) the system control software application (also called NQR GUI) and (2) the embedded system firmware. The general software architecture is illustrated in Fig. 10.2.

The first component is located on the detector's computer and consists of the graphical user interface, the application core, the interface level, and the databases. This is implemented in the Microsoft Visual Studio integrated development environment, using the C# and XAML programming languages, WPF technology and the

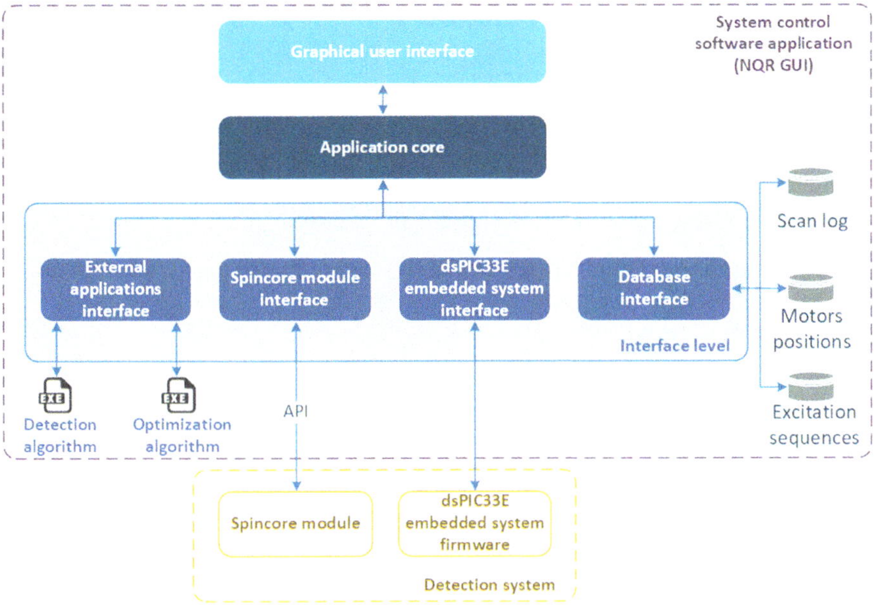

Fig. 10.2 Software architecture of the detection system

.NET 4.6.1 framework. The graphical user interface is the component through which the operator controls the system. It contains several types of controls (buttons, lists, windows, etc.) to provide the user with the functions that the system can perform and to inform him about the status of the detector. The application core is a collection of classes with the role of fulfilling the operator's commands and ensuring the correct operation of the detection system. The interface level consists of several classes that ensure the operation or communication with the various system components, such as:

- detection application (data post-processing), integrating the deep learning algorithm (artificial intelligence);
- excitation sequence optimization application, detailed in [2];
- Spincore module, for pulse sequence generation, detection control, data reception and pre-processing, respectively;
- embedded system for controlling the detection process and monitoring the system;
- databases for storing the scan history, stepper motor positions, and excitation sequences.

The second component is the embedded system's firmware and is responsible for ensuring its operation, e.g., reading the sensors or the reflected wave and ensuring the communication with the application on the computer. The following represent technical aspects regarding the implementation of these software components.

10.2.1 Embedded System Firmware

The embedded system developed with the dsPIC33E microcontroller is involved in carrying out the detection process and ensuring the continuous monitoring of the equipment. The firmware is developed in the C programming language, using the Microchip MPLAB integrated development environment. From a functional point of view, the system meets the following requirements:

- ensuring serial communication with the software application on the computer; the UART port is used, and the symbol rate is 9600 Bd;
- acquisition of the reflected wave;

The shape of the reflected wave is necessary in order to tune the resonant circuit. It is acquired using a 12-bit analog-to-digital converter at a sampling rate of 1 kS/s.

- performing frequency tuning;

The frequency tuning is based on a dedicated algorithm that aims to reduce the average value of the reflected wave by changing the capacitances of the two capacitors.

- stepper motor control;

The control of the stepper motors is necessary for tuning the capacitances in the resonant circuit, in order to achieve the coarse and fine tuning. The controls are transmitted via GPIO ports to the motor control circuit.

- optical conversion of the TX/RX switching signals and fast power amplifier decoupling;

The signals are received from the Spincore module and are converted in order to be transmitted via fiber optic cables to the switch and RF amplifier, respectively. GPIO interrupts were used to implement this function.

- ensuring communication with the power supply unit;

The power supply unit integrates an analog-to-digital converter for reading the voltages generated by it and its temperature, respectively. The I^2C interface was used to provide this function.

- digital attenuator control;

The digital attenuator is used to adjust the signal voltage at the input of the power amplifier (thus, adjusting the gain, given that it has a fixed gain). The SPI port was used for this.

- reading the external temperature sensor;

This is an analog sensor and, therefore, one of the microcontroller's analog ports was used.

- battery charging control;

Batteries charging is used to power the preamplifiers in order to avoid capturing noise by the reception chain during detection. It is controlled via the power supply, using the I^2C port.

- capacitor bank configuration;

The capacitor bank is configured by 5 signals transmitted via fiber optic cables. The GPIO port was used for implementing this function.

10.2.2 Embedded System Interface

The software application on the computer communicates with the embedded system through the serial port. The system interface is implemented as a class (*CommandAndControlModule*) that includes methods related to the functions it performs, such as controlling the motors or reading the reflected wave. The commands that the software sends are encoded characters. E.g., when starting the application, this class calls the connection method, *Connect()*, which reads the embedded system's identifier, using the *GetID()* method, and verifies it.

The *Connect()* method sends the identifier read command to all available ports, waiting for the "Modul comanda" response. Thus, the embedded system can be connected to any serial port of the computer. *Try-catch* blocks are used so that any errors that may occur during the connection process (e.g., unavailable port) do not block the software application. In the *GetID()* method, the identifier read instruction is encoded in the form of the character *i*. Also, since the software application operates on multiple threads, all methods that transmit commands to the embedded system firstly check if there is an ongoing communication.

The communication was implemented using the *SerialCommunication* class and a *SerialPort* object [3], provided by the .NET framework. It includes methods and properties for configuring the serial port parameters (port name, symbol rate, parity, data and stop bits), writing and reading. These functions have been integrated into more complete implementations, which also ensure buffer clearing and setting the variable (flag) which indicates if the communication is ongoing.

Both the *CommandAndControlModule* and *SerialCommunication* are implemented to allow the existence of a single instance (singleton classes). This avoids the risk of instantiating multiple objects, given that the resources it controls are unique (a single embedded system and one secure serial port that it uses for communicating).

10.2.3 Spincore Module Interface

The Spincore module communicates with the software application on the computer via the USB port, using a dedicated API, called *SpinAPI* [4]. The interface is implemented within the *RadioProcessor* class, which contains methods for configuring and controlling (starting/stopping) the pulse sequence, but also for estimating its time. The API provided by the manufacturer is created in the C/C++ programming language and is in the form of a DLL file. There are two ways in which it can be used in the software application:

1. Compiling the Spincore module control program made in C/C++ and calling it as an external executable file;
2. Direct calling of the C/C++ functions from the program made in C#.

The second option was chosen, due to its simplicity, and was implemented by importing the functions from the DLL file and correspondingly declaring the types of parameters and the results provided by them (*marshalling*). Programming the Spincore module consists of the following steps:

- call the *pb_init()* and *pb_set_defaults()* functions to initialize the module;
- use the *pb_core_clock()* and *pb_overflow()* functions to configure the analog-to-digital converter;
- configure the spectral width used for quadrature detection (low pass filter cutoff frequency), with the function *pb_setup_filters()*;
- determine the number of samples to be acquired with *pb_set_num_points()*;
- set the signal amplitude generated by the module to approximately 632 mVpp, equivalent to 0 dBm, using the *pb_set_amp()* function;
- configure the sine and cosine channels used for quadrature detection using *pb_start_programming()* and *pb_set_phase()*;
- set the frequencies used for scanning with the function *pb_set_freq()*;
- define and finalize the excitation sequence using the functions *pb_inst_radio()* and *pb_stop_programming()*;
- start the excitation sequence with the functions *pb_reset()* and *pb_start()*;
- read the acquired data with *pb_get_data()*;
- close communication with the module using *pb_close()*.

The steps presented above are presented in the order in which they were implemented.

10.2.4 Database Interface

The application contains three databases (DB) for storing the pulse sequences, scan history and motor positions. Separate databases were chosen so that they can be easily updated and due to the distinct information they store. The scan history contains

Table 10.1 Structure of the scan history database

Crt. No.	Field name	Data type	Description
1	Id	int	Scan identifier (primary key)
2	ScannedSubstances	nvarchar	List of scanned substances
3	IdentifiedSubstances	nvarchar	List of detected substances
4	ScanDateTime	datetime	Scan date and time

Table 10.2 Structure of the stepper motor position database

Crt. No.	Field name	Data type	Description
1	Id	int	Position identifier (primary key)
2	Label	nvarchar	Position name
3	Frequency	float	Frequency associated with the position
4	Motor1Position	int	Motor position 1 (parallel capacitor)
5	Motor2Position	int	Motor position 2 (series capacitor)
6	RelaysStates	nvarchar	Capacitor bank relays states

all the scans performed, the results and their performance data, and is specific to the equipment. The DB that stores the stepper motor positions contains the step numbers for the two motors corresponding to the coarse tuning positions. The latter are set manually when manufacturing the detector. The DB with excitation sequences contains their parameters and can be updated together with that of the motor positions to add new substances, or independently if only the update of existing sequences is desired.

Databases are implemented using SQL Server Compact 4.0, which is an API and can be integrated into the software application without the need to install a dedicated database engine (for example, SQL Server). Tables 10.1, 10.2 and 10.3 show the structures of the three DBs.

The lists of scanned and detected substances are formed by their comma-separated names. The name of the motors position is a suggestive label, which allows easier identification and is very useful when testing several detection coils. The states of the relays are stored as an array containing the characters *0* (open) or *1* (closed) in the positions associated with the relays, reflecting their states. In the last table, fields 2 and 3 represent the name of the substance associated with the excitation sequence, and two names are stored, because the user interface can be (optionally) displayed in two languages. The excitation power is indirectly indicated by the attenuation value. The last field is used to classify the substance in one of the following categories: explosives, drugs, drugs, toxic and experimental substances. It is used to group detectable substances in the user interface, the latter category being used for research.

The interface with the three DBs is implemented through specific classes: *LogDatabaseManager*, for the scan history, *PositionsDatabaseManager*, for the

Table 10.3 Structure of the excitation sequence database

Crt. No.	Field name	Data type	Description
1	Id	int	Sequence identifier (primary key)
2	NameEnglish	nvarchar	Name of the substance in English
3	NameRomanian	nvarchar	Name of the substance in Romanian
4	Frequency	float	Excitation frequency
5	Teta	float	Width of the excitation pulse (θ_{90°)
6	Tau	float	Time between excitation pulses
7	Scan	int	Number of scans in a sequence
8	Seq	int	Number of sequences
9	T1	float	Longitudinal relaxation time
10	Dpa	float	Time until start of acquisition
11	Da	float	Acquisition window
12	BW	float	Frequency sweep band
13	SW	nvarchar	Spectral width
14	Attenuation	int	Attenuation used for the RF amplifier
15	Category	nvarchar	Substance category

motors positions, and *SequencesDatabaseManager*, for the excitation sequences, respectively. They implement read and write methods based on those provided by the *SqlServerCe* class.

10.2.5 Interface with External Applications

The deep learning and sequence optimization algorithms are compiled as standalone applications, allowing easy updates if they are to be improved or replaced. For simplicity, they were represented in Fig. 10.2 as part of the system control application. The interface with the detection application is implemented by the *MachineLearning* class, which contains methods for opening command line executables, transmitting received data for post-processing and analysis, and receiving the detection result, respectively. The *Process* class [5] is used to call external executables, and the communication with them is done by redirecting their input and output streams and using the *StreamReader* [6] and *StreamWriter* [7] classes. In the case of the optimization application, the interface is done through several defined functions that implement the communication protocol and use a *TcpClient* object.

Post-processing and signal analysis is performed within the external application that contains the deep learning algorithm. The data received from the Spincore module is transmitted to the external application in text format, via the command line. Waiting for the detection result is done on a separate thread, so as not to block the main application while the analysis is performed. The result is indicated on the command line by the text "*True*", in case of confirmation of the presence of the NQR response, respectively "*False*", in case of confirmation of its absence. The implementation of the additional thread was done using the *Thread* class [8], and waiting for the result without loading the processor, using the *AutoResetEvent* class [9]. All these operations are performed in the background, without the operator noticing them.

The excitation sequence optimization application communicates with the external application interface via a TCP connection. It can be internal, with the server and the client located on the computer running the application, if it is used within the NQR system control software application, but it can also remote, for research. During the optimization, a message exchange is performed in which the optimization application provides the excitation sequence that is used to perform the scan, and the control application, through the interface with external applications, responds with the acquired data.

10.2.6 Application Core

The application core is implemented through several classes that ensure the connection between the user interface and the interface level, and ensure the operation of the detection system, respectively. The link classes are *MainWindowViewModel*, *MainWindowResearchViewModel* and *SubstanceDatabaseWindowViewModel*, corresponding to the three windows available in the application. They contain properties for all the objects that are displayed in the windows (lists of substances and sequences, plots, settings, etc.) and methods for implementing the functions available (starting/stopping the detection process, displaying the scan history, etc.). The application core implements its operation according to the diagram in Fig. 10.3.

Upon startup, the application initializes the detection system by starting its modules (power supply control) and performing the initialization test. The latter checks whether the transmission and reception chains are working properly and whether the system voltages have the nominal values. Then it waits for the user's command, which can be, for example, to close the application, view the scan history, change the access mode, select the scanned substances, start the detection process, etc. In the latter case, the scanning process is repeated for each selected substance and consists in performing the tuning (coarse and fine), the excitation, the acquisition and processing of the signal, and the storage of the detection result, respectively. If the application is closed, the motors are first moved to the default (reference) positions before exiting the program.

The initialization and detection processes can take tens of seconds, depending on the performance of the computer running the application, respectively the number of substances chosen for scanning. In order not to block the user interface and to allow the detection process to be stopped at any time, they are executed on separate threads from that of the interface, implemented using the *BackgroundWorker* class [254]. This is similar to the *Thread* class, with the difference that it allows a more simplified

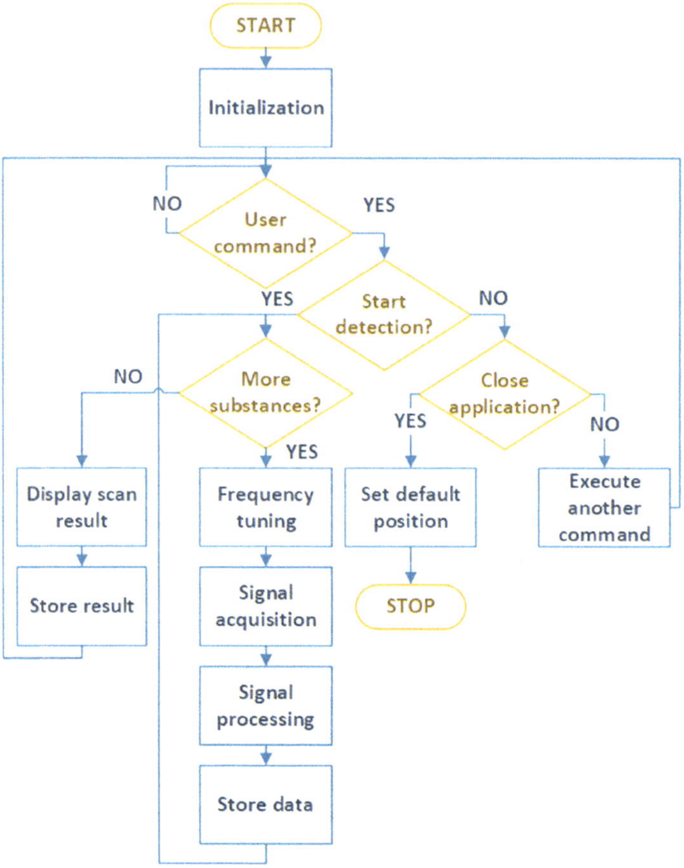

Fig. 10.3 General flowchart of the software application

implementation. The *SystemMananger* class contains methods that implement the initialization and detection processes. It ensures the operations of the detection system in the desired order and in turn calls the interfaces with all their components.

10.2.7 Graphical User Interface

The graphical user interface is implemented using the *MahApps.Metro* framework [10]. It is designed to be easy to use and allows three access modes, called *Operator*, *Administrator* and *Maintenance*. The first mode is used by the system operator and allows access only to the detection function. The second mode also allows access to the scan history, after password authentication. The latter mode also requires

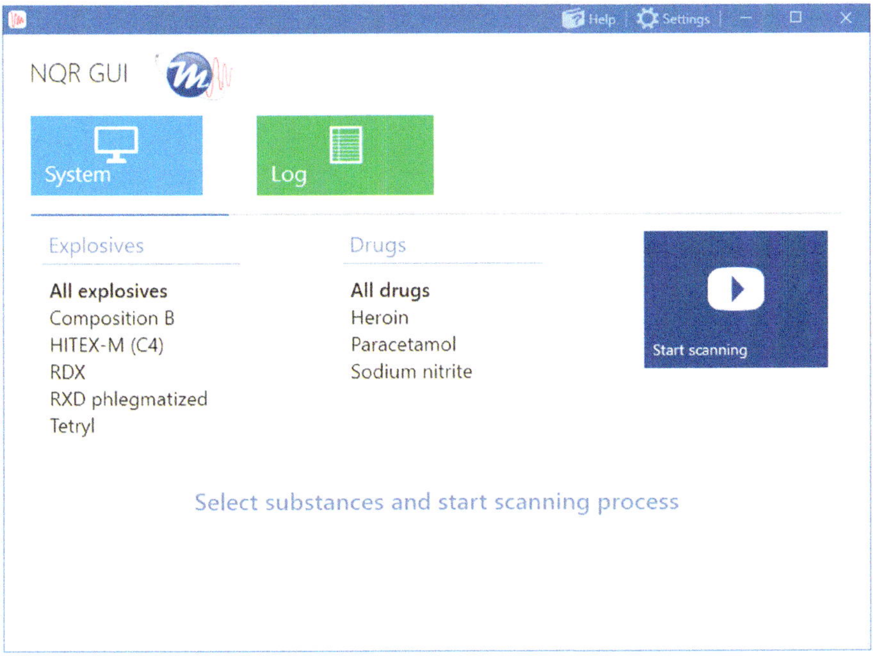

Fig. 10.4 Operator window

password authentication and allows full access to all system functions, such as the substance database, plots, motors control or frequency tuning.

The window associated with the *Operator* mode displays the lists of detectable substances from where the user can choose the substances to be scanned. The *Operator* window is illustrated in Fig. 10.4.

After inserting the object to be scanned, the operator selects either the scan of all substances or the selective scan of those of interest, and initiates the detection process. At the end, the detection result is indicated by a green box if no substance has been detected, respectively red, otherwise. The result of a positive detection is shown in Fig. 10.5. After logging into *Administrator* mode, the user can also view the scan history. The application also allows viewing the scan history in the form of a printable report. In the *Maintenance* mode, the received data can also be viewed, as shown in Fig. 10.6.

The figures illustrate the real and imaginary components after the quadrature detection, the signal in the time (magnitude) and frequency (spectrum) domains, respectively the Lissajous plot (for research purposes). At the bottom, the following statistical information is indicated: average, variance and amplitude of the maximum spectral peak. The *Auto* button in the top bar allows the user to make a configured number of acquisitions and was used to create the data set used to train the detection algorithm. The *Search* button allows the user to start the process of optimizing the excitation sequence.

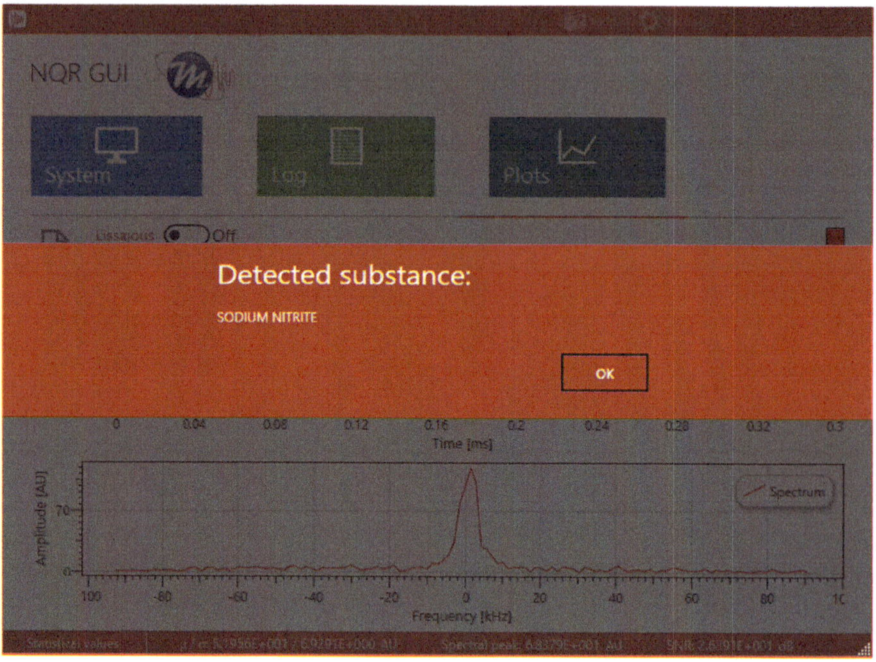

Fig. 10.5 The result of a positive detection in the operator window

The user interface was implemented using the elements provided by the MahApps.Metro framework (buttons, lists, drop-down menus, etc.). The plots were implemented using the Dynamic Data Display library [11], and the reports using WPF Reports [12].

10.2.8 Implementation of the Detection Application

There are two approaches for implementing the detection algorithm: (1) in C# code or (2) in the code in which the algorithm was developed. The first can be done using libraries dedicated to machine learning, such as ML.NET or TensorFlow.NET. The second consists in creating an external application that will be called from the C# code. The first offers the advantage that no external executable should be called, but involves reimplementing and testing the solution in the C# language, with the risk that the accuracy of the detection obtained will not be the same as that achieved in the initial language. Therefore, the second technique was chosen, and in order to avoid the installation on the detector computer of the Python language and machine learning libraries necessary to run the algorithm, the solution was integrated in a command line application and compiled for the Windows operating system. This also avoids the re-implementation errors that would have occurred in the case of the

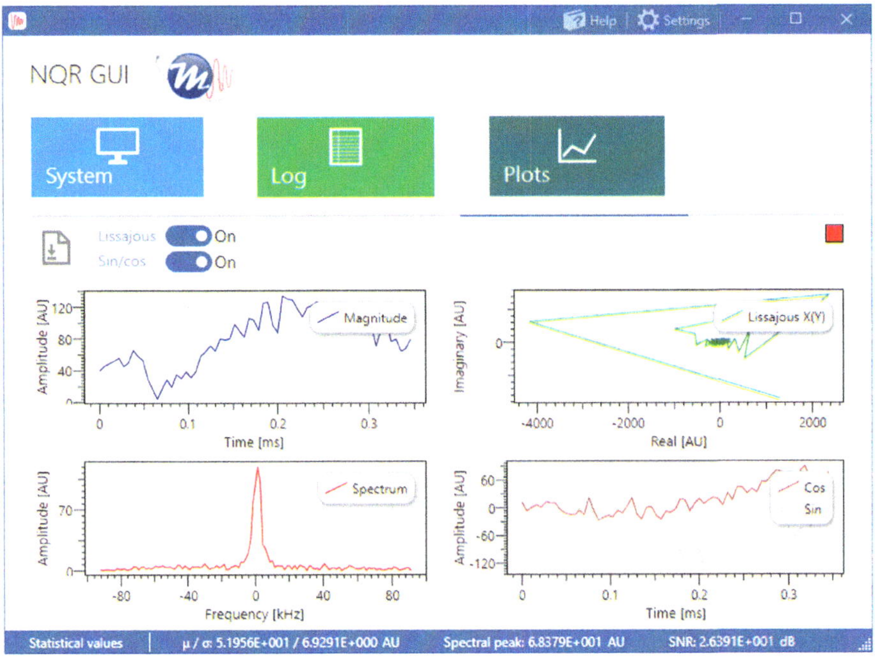

Fig. 10.6 Plots of the received signal

first approach. The solution also has the advantage that it allows easy updates of the algorithm by completely replacing the external application.

The detection application is implemented in the Python language, using the *Cmd* class [13] to create a command line-level interpreter. It implements the program loop where it waits to receive a command and provides the result of its execution. A single command is implemented (*predict*) to perform the detection. The data format expected by the application has been described above. In order to perform the detection, the application uses the *DataLoader* class to create the data structure and the *FeaturesGenerator* class to generate the features based on which the classification and the model itself are created.

The application was compiled using PyInstaller [14] and can run on Windows operating systems without the need to install software libraries. The signal processing and analysis time provided by the Spincore module was measured using the *Stop-Watch* class. It has a value of approx. 3.8 s at the first call of the application, after which it decreases and stabilizes at 1.7 s. Compared to the data from the analysis presented in a previous chapter, the time is between that necessary for statistical algorithms and that of classification solutions, and is at least 2 times shorter than the former, but much longer than the latter. It should be noted that the classification algorithms evaluated in a previous chapter have a much lower complexity than the one implemented and do not reach its detection performance. Also, the time depends on the processor's loading and the processing power of the computer running the algorithm, which can be reduced by using a better system and eliminating non-critical (background) processes in the operating system.

10.3 Evaluation of the Detection System

This section evaluates different characteristics of the component modules, such as the rejection factor of the RF switch, the improvement given by the ferrite core and the pre-processing methods, but also the system's detection limits. Finally, the improvement of the signal-to-noise ratio is evaluated by optimizing the excitation sequences.

10.3.1 Evaluation of the Radio Frequency Switch

The RF switch is used to separate the transmission chain from the reception chain, and the ability to block the reception of the excitation signal by the reception chain is a very important feature. This was evaluated by using the switch in transmission mode, applying to its input the signal taken from an RF synthesizer and measuring the signal at its output. The measurements were performed with the switch separated from the rest of the detection system. The Rohde and Schwarz HM8135 synthesizer was used and several measurements were made in the 0.4–10 MHz frequency band. The results are illustrated in Fig. 10.7.

As seen in the plot, the RF switch has a good noise rejection factor, which varies between 46 and 50 dB in the investigated frequency band.

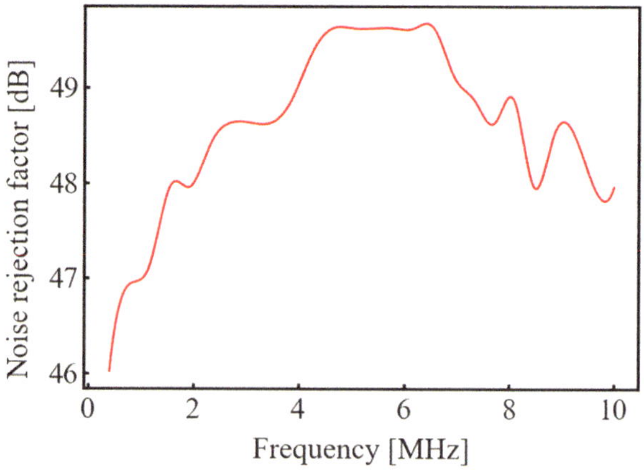

Fig. 10.7 RF switch noise rejection factor versus frequency

10.3.2 Detection Limits

The performance of the detection system was evaluated using 150 g of sodium nitrite and 130 g of paracetamol. The detection limits are evaluated in term of maximum detection distance and the minimum detection limit.

In order to evaluate the maximum detection distance, several data acquisitions were performed with the substance at different distances from the coil, as shown in Fig. 10.8.

As seen, the SNR decreases if the substance is too far from the coil and the response amplitude becomes comparable with the noise. The maximum detection distance can be considered to be 2 cm at an excitation power of 7 W, for paracetamol. A similar analysis is performed for sodium nitrite and the same power of 7 W is used, in order to perform measurements in the same conditions. Since the rotation of the nuclei is proportional to the excitation power [15], the pulse width is experimentally reduced to 110 μs to obtain the optimal excitation. The analysis showed that the maximum detection distance in the case of sodium nitrite is 6 cm at a power of 7 W, indicating that this substance responds more strongly.

In order to evaluate the minimum detection limit, multiple small mass samples were used. These were scanned at the same power of 7 W to be able to compare them with previous results. The limit of detection was identified as 1 g for sodium nitrite and 15 g for paracetamol, according to Fig. 10.9.

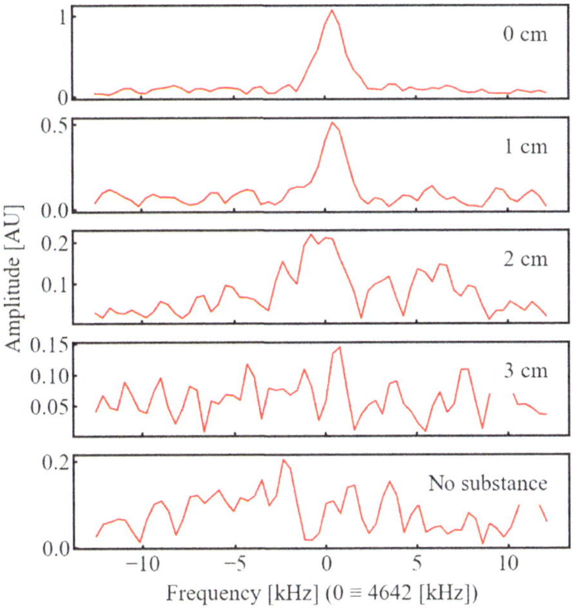

Fig. 10.8 Spectral diagrams obtained after scanning paracetamol at different distances from the coil and in its absence (*AU* arbitrary units)

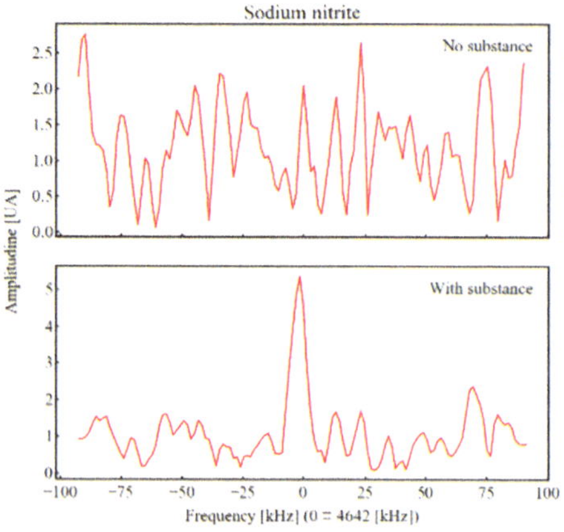

Fig. 10.9 Spectral diagrams obtained in the cases: without substance (top), 1 g of sodium nitrite (bottom left), and 15 g of paracetamol (bottom right) (*AU* arbitrary units)

The substance has a low mass and the NQR response is weaker. As seen from the figures above, the spectral diagrams show a response peak comparable to the noise level. Also, the measured response differs significantly in the substance's absence.

10.3.3 Ferrite Core Improvement

A ferrite core is placed at the center of the detection coil to increase the magnetic field intensity. The improvement of the core is evaluated by measuring and comparing the amplitude of the spectral peak of the sodium nitrite response in both situations: with and without core. Figure 10.10 presents the results obtained.

The spectral diagram shows the slight improvement brought by the ferrite core, increasing the amplitude of the response peak by approximately 2 arbitrary units.

Fig. 10.10 Spectral diagram of the sodium nitrite response in the presence and absence of the coil's ferrite core (*AU* arbitrary units)

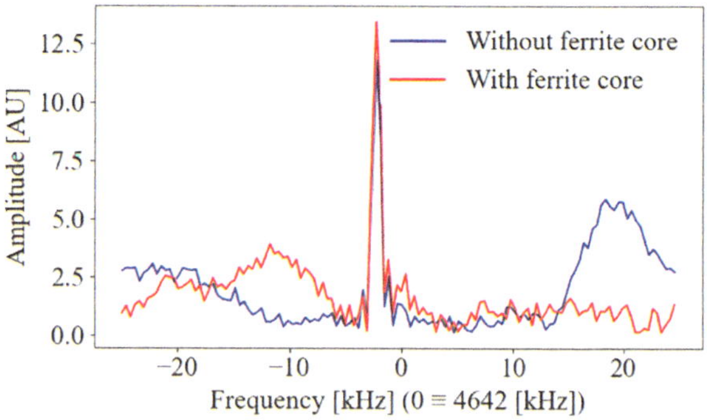

Fig. 10.11 Prototype detection system

10.3.4 Evaluation of the Signal-to-Noise Ratio

The detection system integrates several hardware and software solutions to improve the SNR. To evaluate it, it was necessary to perform measurements on a prototype NQR detector, developed in a previous project [16], and compare with those obtained using the system presented in this work. The prototype detector is illustrated in Fig. 10.11.

The prototype is a predecessor of this system and does not integrate the solutions to improve the detection that were previously described, such as special shielding measures and fiber optic communication, allowing a comparison between its performance and the detector presented, called optimized detector. Several measurements were made in the presence and absence of the substance, using both spectrometers in the same conditions (temperature, excitation sequence, transmitted power, position of the substance relative to the coil). Figures 10.12 and 10.13 show three measurements obtained in each case.

A higher noise level can be observed in the case of the prototype detector, in both situations (in the absence and presence of the substance). The power spectra were calculated and the SNR was estimated from them, by dividing the power of the response peak in the presence of the substance by the power of noise in its absence (average noise), and converting the values to dB. Thus, the following formula was used:

$$\mathrm{SNR} = 20\log_{10}\frac{\max\left(\boldsymbol{y}_{\mathrm{FFT,CS}}\right)^2}{\overline{\boldsymbol{y}}_{\mathrm{FS}}^2} \tag{10.1}$$

where $\boldsymbol{y}_{\mathrm{FFT,CS}}$ represents the discrete Fourier transform obtained in the presence of the substance, and $\overline{\boldsymbol{y}}_{\mathrm{FS}}$ is the average value of the discrete Fourier transform obtained in its absence. The results of the measurements are summarized in Table 10.4.

The values in the table are higher than those in Figs. 10.12 and 10.13, because it represents the powers of the respective signals, which are necessary for the estimation

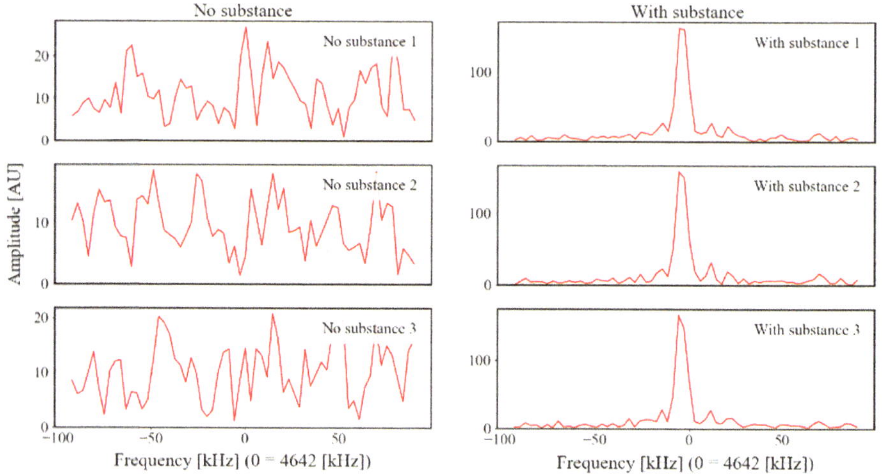

Fig. 10.12 Spectral diagrams obtained after scanning sodium nitrite with the prototype detector, in the absence (left) and presence (right) of the substance (*AU* arbitrary units)

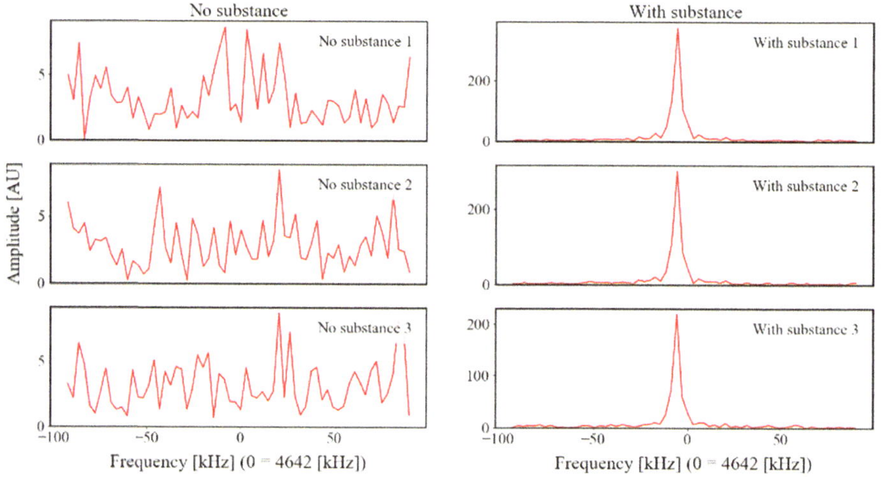

Fig. 10.13 Spectral diagrams obtained after scanning sodium nitrite with the optimized detector, in the absence (left) and presence (right) of the substance (*AU* arbitrary units)

of the SNR. The optimized detector is more efficient, as shown by the increase in SNR by about 31 dB; this is due to the integration of the detection improvement solutions.

In addition, the evaluation of SNR improvement due to pre-processing techniques was performed. Similar measurements were made at different spectral widths to evaluate the contribution of quadrature detection, which, in fact, represent the cutoff frequencies of the low pass filter. Two-coil measurements could not be performed

Table 10.4 Estimates of the signal-to-noise ratio from the prototype and optimized detector

Detection system	Noise spectral power (AU)	Response peak spectral power (AU × 1e3)	SNR (AU)	Average SNR (dB)
Prototype	125.3	27.01	46.66	47.9
	95.3	25.34	48.49	
	104.6	28.05	48.56	
Optimized	10.37	141.8	82.71	78.86
	8.5	91.54	80.64	
	10.42	47.94	73.22	

AU arbitrary units

to show the increase in SNR by $\sqrt{2}$, because the spectrometer is designed with a single detection coil. The spectral widths of 185, 300 and 600 kHz were used, and a slight increase in SNR was observed as the spectral width decreased. The relative differences between the signal-to-noise ratios corresponding to the spectral widths is about 1 dB [17]. As expected, a higher spectral width allows more noise to be picked up by the analog-to-digital converter.

In order to highlight the contribution of the averaging technique, acquisitions were made with the optimized detector in the following conditions: (1) with 100 and (2) with 1000 averages. The average noise was estimated in the absence of the substance and the peak response was measured in both conditions. Figures 10.14 and 10.15

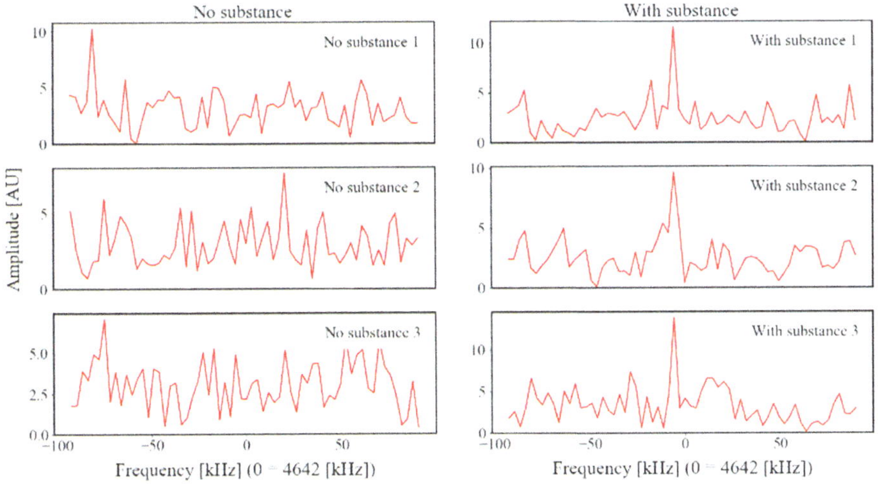

Fig. 10.14 Spectral diagrams obtained after scanning sodium nitrite with the optimized detector, in the absence (left) and presence (right) of the substance, using 100 averages (*AU* arbitrary units) [17]. © 2020 IEEE. Reprinted with permission from Monea [17]

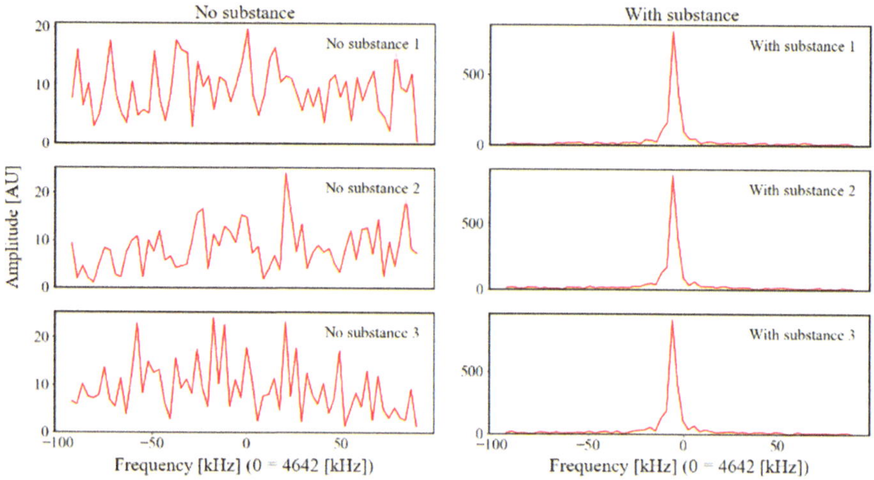

Fig. 10.15 Spectral diagrams obtained after scanning sodium nitrite with the optimized detector, in the absence (left) and presence (right) of the substance, using 1000 averages (*AU* arbitrary units) [17]. © 2020 IEEE. Reprinted with permission from Monea [17]

illustrate three measurements obtained in each case. The results are summarized in Table 10.5.

The increase in the signal-to-noise ratio due to a higher number of averages is obvious and approaches the theoretical increase with cu $\sqrt{10}$, which approximates the ratio between the two SNRs.

The detection system was presented nationally and internationally, and it was awarded a gold medal at the exhibitions in Geneva [18, 19] and Barcelona in 2017 [20]; several patent applications were filed [21–23], and in 2019 the patent was obtained in Romania [24].

Table 10.5 Estimates of the signal-to-noise ratio according to the number of averages

Averages	Noise spectral power (AU)	Response peak spectral power (AU × 1e3)	SNR (AU)	Average SNR (dB)
100	9.57	0.136	23.06	23.36
	8.63	0.093	20.69	
	9.23	0.191	26.34	
1000	9.06	68.39	77.55	79.11
	9.25	78.18	78.53	
	7.54	87.03	81.24	

AU arbitrary units. [17]. © 2020 IEEE. Reprinted with permission from Monea [17]

10.4 Conclusions

This chapter aimed to present the implementation of a system for signal pre-processing, processing and analysis based on the nuclear quadrupole resonance technique.

The detection system developed was described from a hardware point of view. The general block diagram of the detector was presented, after which the component modules were described. The system was designed to scan for explosives, narcotics, toxic substances and drugs in the 0.4–6 MHz frequency band. Several noise reduction solutions have been proposed and implemented, such as the use of fiber optics for the transmission of control signals to modules near the analog circuits and powering the low noise amplifiers from batteries that are electrically isolated from the rest of the system. In addition, the system integrates a capacitors bank to ensure frequency tuning throughout the working band, by adding to the detection circuit an adjustable capacitance of up to 29.5 nF. The detector allows the scanning of 8 substances using multi-pulse type excitation sequences, with frequency sweeping and adjusted to each substance.

The software description of the detection system presented the general architecture and its software components. The operation of the detector is performed by the high-level software application, located on the computer, called the system control application. It is implemented in the C# language and integrates the user interface, respectively the classes that ensure its connection to external deep learning and optimization applications, databases and system modules. There are three levels of access for the user, through which he can start or stop the detection process, view the scan history and plots, or even have access to all system functions. The low-level application, implemented in the C language, is the firmware of the embedded system that ensures the connection between the application on the computer and the detector modules, such as temperature sensors, voltage source, stepper motors, etc. The detection and optimization algorithms have been implemented as stand-alone Python applications that are called through the command line by the system control application. They have been compiled so that they contain all the components necessary for their operation, and the solution has the advantages that it does not require the installation of software libraries on the computer on which they are deployed and they can be easily updated. The measured processing and analysis time was about 1.7 s and can be reduced by using a more powerful computer, respectively optimizing the processor load. The optimization application implements a protocol that ensures communication with the system control application, and, at each scan, sends the excitation sequence containing the new set of parameters to the control, waits for the measurement, then receives and processes the acquired data.

Finally, the detection system was evaluated on sodium nitrite and paracetamol, measuring the distance and the substance quantity according to the excitation power. It was shown that at an excitation power of 7 W, the maximum detection distance was 2 cm for paracetamol and 6 cm for sodium nitrite. The minimum limit of detection was 1 g of sodium nitrite and 15 g of paracetamol, using the same power. Also, the solutions to improve the detection have increased the signal-to-noise ratio by approximately 31 dB compared to a prototype detector that does not integrate such solutions.

References

1. Monea, C., Iana, G.V., Ionita, S., Ionescu, L.M., Zaharia, S.A., Ilie, S., Bizon, N.: An optimized NQR spectrometer for detection of prohibited substances. Measurement **151**, 107158 (2020)
2. Monea, C.: Optimization of NQR excitation sequences using black-box techniques. J. Magn. Reson. **321**, 106858 (2020)
3. Microsoft Docs: SerialPort Class, 20 Aug 2020. Internet: https://docs.microsoft.com/en-us/dotnet/api/system.io.ports.serialport?view=netframework-4.8
4. SpinCore Technologies: SpinAPI Documentation, 20 Aug 2020. Internet: http://www.spincore.com/support/spinapi/reference/production/2013-09-25/index.html
5. Microsoft Docs: Process Class, 20 Aug 2020. Internet: https://docs.microsoft.com/en-us/dotnet/api/system.diagnostics.process?view=netframework-4.8
6. Microsoft Docs: StreamReader Class, 20 Aug 2020. Internet: https://docs.microsoft.com/en-us/dotnet/api/system.io.streamreader?view=netframework-4.8
7. Microsoft Docs: StreamWriter Class, 20 Aug 2020. Internet: https://docs.microsoft.com/en-us/dotnet/api/system.io.streamwriter?view=netframework-4.8
8. Microsoft Docs: Thread Class, 20 Aug 2020. Internet: https://docs.microsoft.com/en-us/dotnet/api/system.threading.thread?view=netframework-4.8
9. Microsoft Docs: AutoResetEvent Class, 20 Aug 2020. Internet: https://docs.microsoft.com/en-us/dotnet/api/system.threading.autoresetevent?view=netframework-4.8
10. MahApps.Metro: 20 Aug 2020. Internet: https://mahapps.com/
11. CodePlex Archive: Dynamic Data Display, 20 Aug 2020. Internet: https://archive.codeplex.com/?p=dynamicdatadisplay
12. CodePlex Archive: wpfreports, 20 Aug 2020. Internet: https://archive.codeplex.com/?p=wpfreports
13. Python 3.8.2 Documentation: cmd—Support for Line-Oriented Command Interpreters, 20 Aug 2020. Internet: https://docs.python.org/3/library/cmd.html
14. PyInstaller: 20 Aug 2020. Internet: https://www.pyinstaller.org/
15. Cardona, L.R.: Nuclear Quadrupole Resonance System for Landmine Detection in Antioquia, Ph.D. thesis, National University of Colombia (2017)
16. Mira Technologies Group: The use of nuclear quadruple resonance method to produce equipment for detection of drugs and explosives—DESMTEH (in Romanian language: "Utilizarea metodei rezonanței quadripolare nucleare la producerea de echipamente pentru detecția explozivilor și drogurilor—DESMTEH"), national research project, POS CCE programme
17. Monea, C.: Study of signal pre-processing techniques in nuclear quadrupole resonance spectroscopy. In: 11th International Conference on Electronics, Computers and Artificial Intelligence (2020)
18. RFI Romania: The NQRScan detector, awarded at the International Exhibition of Inventions of Geneva (in Romanian language: "Detectorul NQRScan, premiat la Salonul Internațional de la Geneva"), 24 Aug 2020. Internet: https://www.rfi.ro/eveniment-94253-detectorul-nqrscan-premiat-la-salonul-international-de-la-geneva
19. Mira Technologies Group: The NQRScan detector (in Romanian language: "Detectorul NQRScan"), winner of the International Exhibition of Inventions of Geneva 2017, 24 Aug 2020. Internet: http://www.miratechnologies.ro/premii/
20. Mira Technologies Group: The NQRScan detector (in Romanian language: "Detectorul NQRScan"), winner of the International Exhibition of Inventions INNOVA 2017, 24 Aug 2020. Internet: http://www.miratechnologies.ro/premii/

21. Zaharia, S., Apostol, M., Ionita, S., Iana, V., Monea, C., Ionescu, L.M., Anghel, D., Ilie, M., Varga, A.: Fix detector for detecting substances with explosion hazard, explosives and drugs in luggage, based on nuclear quadrupole resonance (NQR) (in Romanian language: "Detector fix pentru descoperirea substanțelor cu risc exploziv, a explozivilor și a drogurilor din bagaje pe baza efectului rezonanței nucleare cuadrupolare (NQR)"), OSIM patent, no. RO131586 (A0), 30 Dec 2016
22. Zaharia, S., Apostol, M., Ionita, S., Iana, V., Monea, C., Ionescu, L.M., Anghel, D., Ilie, M., Varga, A.: Mobile detector and method for detecting potentially explosive substances, explosives and drugs by nuclear quadrupole resonance (NQR), European patent application, no. WO2018124905 (A1), 05 July 2018
23. Zaharia, S., Apostol, M., Ionita, S., Iana, V., Monea, C., Ionescu, L.M., Anghel, D., Ilie, M., Varga, A.: Mobile detector and method for detecting potentially explosive substances, explosives and drugs by nuclear quadrupole resonance (NQR), USA patent, no. US20200182810 (B2), 16 Feb 2021
24. Zaharia, S., Apostol, M., Ionita, S., Iana, V., Monea, C., Ionescu, L.M., Anghel, D., Ilie, M., Varga, A.: Mobile detector and detection method for explosive substances, explosives and drugs, based on the nuclear quadrupole resonance effect (NQR) (in Romanian language: "Detector mobil și metodă de detecție a substanțelor cu risc exploziv, a explozivilor și a drogurilor pe baza efectului rezonanței nucleare cuadrupolare (NQR)"), OSIM patent, no. RO131585 (B1), 30 Apr 2019

The manufacturer's authorised representative in the EU is Springer Nature Customer Service Centre GmbH, Europaplatz 3, 69115 Heidelberg, Germany. If you have any concerns regarding our products, please contact ProductSafety@springernature.com

Printed and bound by CPI Group (UK) Ltd, Croydon, CR0 4YY
30/09/2026
02225610-0007